Writing an Interpreter in Object Pascal:

Part 2: The Runtime System

Herbert M. Sauro
Seattle, WA

Ambrosius Publishing

First Edition, version 1.01
Published by Ambrosius Publishing and Future Skill Software
books.analogmachine.org

Typeset using LaTeX 2_ε, TikZ, PGFPlots, WinEdt,
and 11pt Math Time Professional 2 Fonts

pgf version is: 3.1.5b

pgfplots version is 1.16

ISBN 10: 1-7325486-2-5 (paperback)
ISBN 13: 978-1-7325486-2-6 (paperback)

Printed in the United States of America.

Mosaic image modified from Daniel Steger's Tikz image (http://www.texample.net/tikz/examples/mosaic-from-pompeii/

If you've purchased the printed copy you can get a 50% discount for the pdf version using the following code:

PDF Discount code: cambrian$
at https://gum.co/oYxzK

Contents

Preface

This is Part 2 of a series I am writing on how to write interpreters in Object Pascal. Part 1, published in early 2019, described the construction of a tokenizer and syntax checker for a simple procedural language I named after my dog, Rhodus. Part 1 ended with a simple calculator application and a full language recognizer.

In Part 2, we will cover the building of the first version of a virtual machine together with a simple assembler. The book then describes combining the virtual machine with the syntax checker for the Rhodus language. In Part 2 we generate the code directly as we parse the source code. In Part 3 of the series, we will separate parsing from code generation via an abstract syntax tree. At the end of Part 3, there will be a very serviceable interpreter that supports user-defined functions, built-in functions, string, and list support together with a range of looping constructs. The big omission from this version is support for libraries, that will come in Part 3.

There are many people and organizations whom I should thank, but foremost must be my infinitely patient wife, Holly, and my two boys Theodore and Tyler, who have put up with the many hours I have spent working alone. I want to thank Holly, in particular, for helping me edit the text. Naturally, I am responsible for the remaining errors, or as a contributor (Marc Claesen) to StackOverflow once humorously remarked, 'Making the manuscript error-free is left as an exercise for the reader.'.

Many thanks to the authors of the TEX system, MikTeX (2.9), TikZ (3.1.5b), PGF-Plots (1.16), WinEdt (10.2), and Affinity Designer (1.8), for making available such amazing tools for technical authors. It is these tools that make it possible for individuals like myself to publish. Also, not forgetting companies such as Createspace/KDP, and Ingramspark that make independent publishing possible at affordable prices. Finally, I should thank Michael Corral (`http://www.mecmath.net/`) and Mike Hucka (`www.sbml.org`), whose LATEX work inspired some of the styles I used in the text.

All code can be obtained from: `https://github.com/penavon/BookPart2`

February 2020
Seattle, WA

HERBERT M. SAURO

Source Code

The source code can be obtained from GitHub at:

`https://github.com/penavon/BookPart2`

1

Interpreters

1.1 Introduction

Welcome to Part 2 of the book series on writing an interpreter using Object Pascal. In Part 1 we looked at how to tokenize source code into tokens, how to parse those tokens for syntactical structure, and how to build a simple evaluator of infix expressions. Along the way, unit testing was introduced via DUnitX, and we covered some of the essential concepts in parser theory.

At the end of Part 1, a grammar for the first version of a language named Rhodus and an application that could recognize and check a script written in the language were presented.

Let's recall the full syntax specification for Rhodus version one that was given in the last Chapter in Part 1 of the book:

```
mainProgram          = statementList [ ';' ] endOfStream
statementList        = statement { [ ';' ] statement }
statement            = assignment | forStatement | ifStatement
                       | whileStatement | repeatStatement
                       | returnStatment | breakStatement
                       | function | endOfStream
variable             = simpleVariable | indexVariable
list                 = '{' [ expressionList ] '}'
simpleVariable       = identifier
indexVariable        = identifier '[' expressionList ']'
expressionList       = expression { ',' expression }
assignment           = variable '=' expression
```

```
function              = FUNCTION identifier '(' [ argumentList ] ')' functionBody
functionBody          = statementList END
argumentList          = argument { ',' argument }
argument              = identifier | REF variable
returnStatement       = RETURN expression
breakStatement        = BREAK
relationOpExpression = simpleExpression
                            | simpleExpression relationalOp simpleExpression
expression            = relationalExpression
                            | relationalExpression BooleanOp relationalExpression
simpleExpression      = term { addingOp term }
term                  = factor { multiplyOp factor }
power                 = { '+' | '-'} factor [ '^' power ]
factor                = '(' expression ')' | variable | number | string |
                             NOT factor | list | functionCall
functionCall          = identifier '(' [ expressionList ] ')'
addingOp              = '+' | '-'
multiplyOp            = '*' | '/' | MOD | DIV
relationalOp          = '==' | '!=' | '<' | '<=' | '>=' | '>'
BooleanOp             = OR | AND | XOR
whileStatement        = WHILE expression DO statementList END
repeatStatement       = REPEAT statementList UNTIL expression
forStatement          = FOR identifier = forList DO statementList END
forList               = value TO value | value DOWNTO value
ifStatement           = IF expression THEN statementList ifEnd
ifEnd                 = END | ELSE statementList END
```

Everything in square brackets is optional, and the vertical line represents 'or'. I've made some changes to the grammar since Part 1. These include a change to avoid a pitfall when using Boolean and relational operators. The new syntax now follows C rules for precedence for Boolean and relational operators. I've introduced two new grammar rules, `relationOpExpression`, and `BooleanOp`. Previously the grammar followed Pascal precedence rules. The new grammar means that `OR,` `AND`, and `XOR` have higher precedence than the relational operators. This means an expression such as `a > 4 and b > 2` is valid and doesn't have to be written as `(a > 4) and (b > 2)`.

I've also changed slightly how semicolons are dealt with. The only place semicolons appear in the grammar are in the first two rules:

```
mainProgram        = statementList [ ';' ] endOfStream
statementList      = statement { [ ';' ] statement }
```

In the latest version I've made both semicolons optional. This means the following two lines are valid syntax:

```
a = 1 b = 2 c = 3
```

```
a = 1; b = 2; c = 3
```

I've not seen any unwanted side effects of this change, and all the tests pass.

Some examples of Rhodus scripts are given below. The first example shows a user-defined function to implement a binary search to look for `value` in a `list`. Note that in version one of the language, there is no way to find the length of a list; hence we pass the length as the second argument.

```
// Binary Search
function doBinarySearch(list, length, value)
    low = 0;
    high = length;
    while 0 <= high do
        mid = (low+high) div 2;
        if list[mid] > value then
           high = mid - 1
        else
           if list[mid] < value then
              low = mid + 1
          else
             return mid
          end;
        end;
    end;
    return -1;
end;

print (doBinarySearch ({3,5,2,8}, 4, 8));
```

A second example shows a bubble sort:

```
// Implement a bubble sort on a list of values
function bubbleSort (list, length)
  sorted = False;
  while not sorted do
       sorted = True;
       length = length - 1;
       for i = 0 to length - 1 do
         if list[i+1] < list[i] then
           temp = list[i];
           list[i] = list[i + 1];
           list[i + 1] = temp;
           sorted = False
         end;
       end;
  end;
end;
```

```
l = {4,2,7,6};
bubbleSort (l, 4);
println (l);
```

An example of a recursive function to compute the Fibonacci series is given below:

```
function Fibonacci(value)
  if value < 2 then
    return value
  else
    return Fibonacci(value - 1) + Fibonacci(value - 2);
end
```

In subsequent volumes of the series, the language will be expanded to include modules, homogenous arrays, exception handling, together with standard libraries for input/output, lists, and strings, arrays, as well as other features.

1.2 Components of an Interpreter

At the most basic level, an interpreter will read a script of instructions and execute them. There are various ways this can be accomplished; for example, the interpreter can go line by line executing the instructions on each line. Early versions of BASIC used this technique. The disadvantage is that in a loop, one will be continually and repeatedly decoding the instructions as there is no record of what instructions were interpreted in the past. The advantage is that such interpreters are relatively easy to write. More advanced interpreters, also developed early on in the history of software, convert a programming language into lower level code that is more easily executed. The advantage is that the original source code only needs to be decoded once; after that, the application executes the simpler code. The simpler code is often called intermediate code or virtual machine code. The application that executes the intermediate code is called the **virtual machine**. The conversion of the high-level source code into intermediate code is called compilation. Figure 1.1 shows a high-level view of these stages.

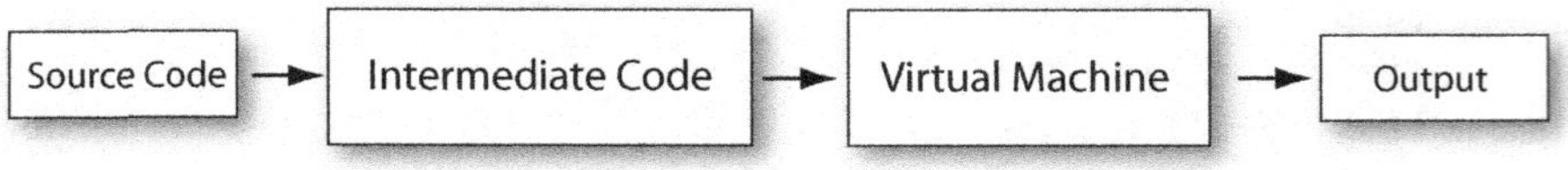

Figure 1.1 Simplified flow from source code to output in an interpreter.

In Part 2 of the series. We will be concerned with the construction and operation of the virtual machine as well as combining the syntax stage with the generation of virtual machine instructions. At the end of Part 2, we will have a functional and capable interpreter.

1.3 FastMM4

Whenever we write software in compiled programming languages such as C, Object Pascal, etc. it is often unavoidable that we'll be creating objects, and allocating dynamic memory. In such cases, we are faced with the task of making sure we don't leak memory; that is, we free up all memory that we no longer need. If the source code is straightforward, the use of **try/finally** blocks is sufficient to free up objects or other resources that are no longer required. A common pattern we find is shown below:

```
obj := TMyObject.Create;
try
   .. do things with obj ..
finally
  obj.Free;
end;
```

Problems arise when an application becomes more complex with multiple interacting objects, complex data structures, spanning multiple source code units. In these instances, it is very easy to forget to free all unused memory. In writing an interpreter, there will be many opportunities where its possible to leak memory. The question is, what tools can we use to help track such leaks? By far, the most useful tool you can use is FastMM4. FastMM4 is a memory manager that was introduced with Delphi 2006. The advantage of using FastMM4 is that it can track your memory allocations, and when the application exits, it can report what memory still remains unfreed.

I used FastMM4 extensively in writing the interpreter, and it helped me track down both obvious and not so obvious memory leaks. Using FastMM4 is straightforward. The version that comes with Delphi is a cut-down edition of the full version. The full version adds additional debugging features, such as a stack trace. To use the cut-down version add the following line to the main program:

```
ReportMemoryLeaksOnShutdown := True;
```

If you run the debug version of your program and there is a leak, you'll get a report at the end. However, you need to run the application from a terminal window because any messages about leaks will quickly vanish if you run the appetition from the Delphi IDE itself. This happens even if you add a **readln** at the end because output from FastMM4 occurs afterward. Here is an example of a leak reported using the FastMM4 that comes with Delphi:

```
Unexpected Memory Leak
An unexpected memory leak has occurred. The unexpected small block leaks are:

13 - 20 bytes: Unknown x 1
69 - 76 bytes: TStringList x 1
```

I've found that there are enough clues in this fairly basic report to fix many of the errors. However, if you're after more information, you need to use the full version of FastMM4. To do that, you need to first download the full version from https://github.com/pleriche/FastMM4/releases. You'll need to copy the source files from the full version to a suitable place where they can be found by the compiler. I have them in a folder called `library`, which is at the same level as all my other projects. Inside `library`, I have the FastMM4 folder.

Next, add FastMM4 as the first unit to include in your project. In my Project dpr unit, I have:

```
uses
{$IFDEF DEBUG}
  FastMM4 in '..\..\Library\FastMM\FastMM4.pas',
{$ENDIF}
  Windows,
  etc.
```

The debug flag allows one to easily generate release versions of your application that don't have FastM4 present.

The last thing to do is to make sure the debug dll, `FastMM_FullDebugMode.dll` is accessible. You'll find a copy of this dll (including the 64-bit version) in:

`\FastMM\FullDebugModeDLL\Precompiled`

You can copy this to the Delphi install folder in bin, for example:

`C:\ProgramFiles(x86)\Embarcadero\Studio\20.0\bin`

or you could even copy it to where the application exe will be. And that's all you have to do. There is an options file called `FastMM4Options.inc`, which you'll find in the FastMM folder. This file has many options that can be changed to alter the behavior of FastMM4, but the defaults seem adequate.

1.4 Useful Reading

1.4.1 Introductory Books

1. Ball, Thorsten. Writing A Compiler In Go. Thorsten Ball, 2018.

2. Kernighan, Brian W.; Pike, Rob (1984). The Unix Programming Environment. Prentice-Hall. ISBN 0-13-937681-X.

3. Nisan, Noam, and Shimon Schocken. The elements of computing systems: building a modern computer from first principles. MIT press, 2005.

4. Parr, Terence. Language implementation patterns: create your own domain-specific and general programming languages. Pragmatic Bookshelf, 2009.

1.4.2 More Advanced Books

1. Jim Smith, Ravi Nair, Virtual Machines: Versatile Platforms for Systems and Processes, Morgan Kauffmann, June 2005

2. Aho, Alfred V., Ravi Sethi, and Jeffrey D. Ullman. Compilers: Principles, Techniques and Tools (also known as The Red Dragon Book), 1986.

1.4.3 Source Code

1. Mak, Ronald. Writing compilers and interpreters: an applied approach/by Ronald Mark. 1991

Note, this is the first edition, 1991. The code is in C, which I found to be understandable. The later editions that use C++ are not as clear. The issue I found is that the object orientated approach that's used tends to obscure the design principles of the interpreter and requires much study to decipher, The C version is much more straightforward.

2. Wren: `https://github.com/wren-lang/wren`.

Of the open source interpreters on GitHub, I found this to be the easiest to read. It's written by Bob Nystrom in C, the same person who is writing the web book: Crafting Interpreters `https://craftinginterpreters.com/`.

3. Gravity: Another open source interpreter worth looking at is Gravity (`https://github.com/marcobambini/gravity`). Gravity, like Wren, is also written in C.

4. If you prefer Go, then the source code to look at is the interpreter written by Thorsten Ball (see book reference above).

There are umpteen BASIC interpreters and other languages that can be studied.

2
The Virtual Machine

2.1 Introduction

Figure 2.1 is the simplified diagram of an interpreter we saw in Chapter 1. It illustrates the generation of intermediate code from source code and execution of the intermediate code by a virtual machine. In this chapter, we will look closely at the operation and implementation of the virtual machine.

Figure 2.1 Simplified flow from source code to output in an interpreter.

2.2 The Virtual Machine

The key concept in an interpreter is the conversion of source code into something much easier to execute. A traditional compiler converts source code to raw machine code, such as Intel x86 binary code or ARM instruction code. This is the simplest code one can write to because it can actually be executed by the hardware directly. This allows programs to run at their maximum speed. Modern interpreters, on the other hand, generate machine code for a fictitious computer, called a **virtual machine**. For example, consider the following line of source code, an assignment, and a simple computation:

```
a = x + 1
```

This source code might be translated to virtual machine code such as:[1]

```
LOAD  x     Load the value of x into the accumulator
ADD   1     Add one to the accumulator
STORE a     Store the value of the accumulator to a
```

A significant advantage of writing code for a virtual machine is that it becomes much easier to port interpreters to different hardware platforms, and perhaps even more important, it allows us much finer control over the execution environment. The virtual machine code is much easier to execute than the original high-level language from which it was derived.

A virtual machine is responsible for executing virtual machine code. The phrase virtual machine refers to the fact that the virtual machine doesn't exist as a piece of physical hardware but is emulated via a software layer.

The concept of a virtual machine goes back at least to the 1960s, where programming languages such as SNOBOL and Lisp used virtual machines to port these languages to different computers; the idea being that it is easier to recompile the virtual machine than write specific code generators for each computer architecture.

In the 1960s, we also saw the development of an interpreter based programming language called BASIC. However, many BASIC interpreters, even in the 1980s, didn't produce virtual machine code that would be later executed but interpreted each line of source code directly, sometimes these are called syntax directed interpreters[2], though the terminology differs between authors. This saved memory space, which was especially important to the early microprocessor-based computers in the 1970s and early 80s. This design was, however, inefficient because within a loop, for example, the source code would have to be repeatedly reinterpreted. Today most BASIC interpreters use the virtual machine model.

One of the most well known early virtual machines was the p-code machine (https://en.wikipedia.org/wiki/P-code_machine), popularised by the UCSD Pascal implementation that executed p-code. An example of p-code for executing `a = x + 1` is shown below:

```
lda  a   load address of a on to the stack
lod  x   load value of x on to the stack
ldc  1   load constant 1 onto the stack
adi      add the top two items on the stack and push result onto the stack
sto      Pop value off stack followed by the address of a; store value to a
```

[1]This example is modeled after the language CESIL: https://en.wikipedia.org/wiki/CESIL

[2]Language Implementation Patterns, Terrance Parr 2010

p-code uses the concept of a stack, which we will return to shortly. The use of virtual machines that execute an intermediate language gained considerable popularity in the 1980s and 90s with the development of scripting languages such as Matlab, Python, Ruby, or R as well as languages such as SmallTalk-80. The technology advanced further with the development of Java (1990s) and the .NET runtime system in 2001.

The very popular programming language Python is based on a virtual machine model. Here is the virtual machine code generated by Python for the statement `a = x + 1`:

```
LOAD_GLOBAL      x
LOAD_GLOBAL      1
BINARY_ADD
STORE_FAST       a
```

The machine code examples we've shown in CESIL, p-code, and Python look very similar. This isn't surprising since they are emulating real hardware, which is based on the idea of registers, operations on registers, and transfer of data to and from registers and main memory.

The main downside of using a virtual machine is the inevitable slowdown in performance when compared to raw machine code generated by a compiler. However, interpreted languages tend to be more easily portable to different computer architectures. Moreover, because everything is run within a virtual machine, it is much easier to control the execution environment. This will invariably prevent the common memory errors we see in applications generated via a compiler. A virtual machine creates a protected sandbox for executing application scripts. This is one reason why interactive computing environments such as Python use virtual machines in their implementation because the user is protected from the common and often fatal errors that can happen when running compiled applications. Interpreted languages also tend to be interactive, which is another significant benefit for users.

Emulating Hardware

There are several ways a virtual machine can emulate hardware. The most common include register and stack-based systems. A register-based virtual machine closely models the registers found in an actual microprocessor. A stack-based virtual machine doesn't emulate registers but creates a stack-based data structure, and all operations are applied to the contents of the stack (See Box 1.1).

Of the two, stack-based virtual machines are much easier to write, and this is what we will focus on in this book. A study in 2005 (Shi et al., Virtual Machine Showdown: Stack Versus Registers) showed that register-based virtual machines are faster by about 30% but with added complexity on managing register allocations.

Box 1.1 Stack-Based Data Structure

A stack is a common data structure found in computer science. It is an ordered list of data elements where elements can be added or removed at one end of the list, called the top of the stack. Imagine elements of the list indexed from zero upwards. Initially, the stack is empty, and we set the value that points to the top of the stack to -1. As items are added to the stack, the value for the top of the stack increases. If we remove items from the top, the value for the top of the stack decreases. The operation that puts items onto the stack is called `push`, while the operation that removes items is called `pop`. A characteristic of a stack is that the last item added is always the first item to be removed, so-called LIFO (last in, first out).

A stack-based virtual machine will have the following elements (Figure 2.2):

1. A stack for storing intermediate results.
2. An instruction pointer (IP) that indicates the location of the next instruction.
3. One or more symbol tables for persistent storage of variables.
4. An instruction handler that:
   - Fetches the next instruction indicated by the instruction pointer.
   - Decodes the instruction.

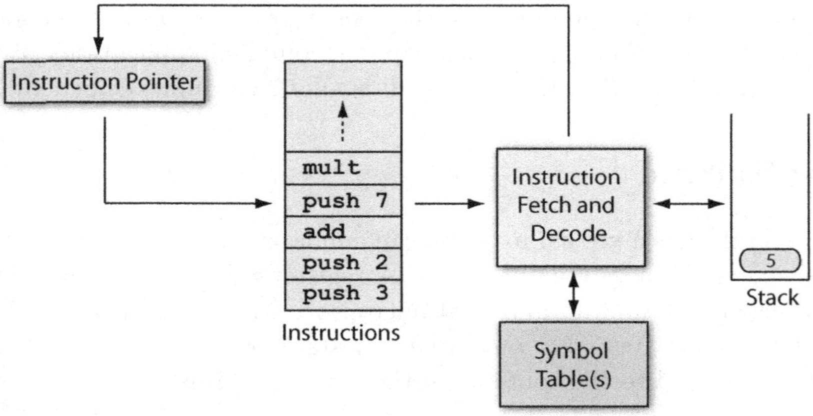

Figure 2.2 Structure of a stack-based virtual machine.

Bytecode

Traditionally, machine code instructions for virtual machines are represented as bytes. For example, both p-code and Java bytecode use bytes to represent machine instructions. Given that a byte is eight bits, it means that a total of 256 different instructions can be encoded using a byte code. This is generally more than enough. For example, Java has 202 byte codes, with the remaining 54 codes reserved for future use[3]. The common intermediate language used by .NET has 234 instructions[4]. In the design of our own virtual machine, we will also use a byte to represent an instruction.

For those instructions that require an argument, we will encode the argument using a 32-bit word. Figure 2.3 illustrates the structure of a single machine instruction that we will use in our virtual machine.

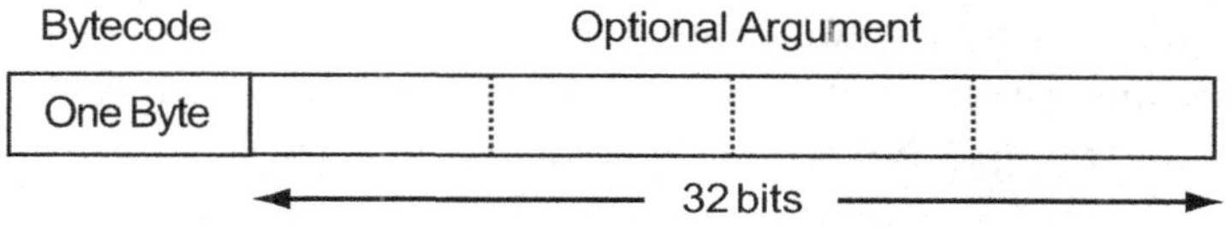

Figure 2.3 Structure of a machine instruction: bytecode followed by an optional 32-bit argument.

This means that a full machine instruction can be up to five bytes long. A side effect of this is that a single instruction won't align on a 4-byte boundary, which could reduce performance. It is possible to pad an instruction with an additional three bytes so that the size grows to eight bytes, which has better alignment properties. If the bytecode is stored to disc, it is possible to strip off the padding to make the code more compact, and when read back in, the padding can be added. This is an implementation detail that we can return to later.

2.3 Stack-based Machines

Some of you may be old enough to remember reverse polish notation (RPN) based calculators. Such calculators can still be purchased today but were at one time much more popular. In the 1970s and 80s, Hewlett-Packard seems to have used RPN in all their calculators.

To explain RPN, let's first consider the commonly used algebraic notation we learn in high school called **infix**. In infix, we generally place the operators (such as

[3] https://en.wikipedia.org/wiki/Java_bytecode
[4] https://en.wikipedia.org/wiki/List_of_CIL_instructions

addition or multiplication) between the two operands. For example, the expression $1 \mathcal{C} 2$ has two operands, 1 and 2, separated by the addition operator. One limitation of infix is knowing which operators to apply first if we have a sequence of operations to compute. For example, $1 \mathcal{C} 2 \times 3$ is ambiguous because do we compute the addition first or the multiplication? The order makes a difference to the final result. There are two solutions to this. We either designate operator precedence; for example, we state that multiplication must always be done before addition, or we can use parentheses to group operations that need to be carried out first. For example, we might write $\triangleright 1 \mathcal{C} 2 \triangleleft \times 3$, indicating that in this case, addition comes first. Operator precedence and bracketing, although handled easily by humans, can be awkward to deal with by computer. This is because the computer has to look ahead to check if there are parts of an expression that should be executed first. As a result, other ways have been devised to express expressions which are simpler to evaluate by computer.

In addition to infix, there are two other ways to express algebraic equations; these include **postfix** and **prefix** notation. In postfix notation, which is of particular interest to us, the two operands come first followed by the operator, for example:

$$\boxed{1\ 2\ \mathcal{C}} \text{ represents } \boxed{1\ \mathcal{C}\ 2} \text{ in infix}$$

For the earlier example $1 \mathcal{C} 2 \times 3$, we would write the expression in postfix as:

$$\boxed{1\ 2\ \mathcal{C}\ 3\ \times}$$

The expression reads as follows: add 1 and 2 together. Use the result to multiply by 3. Notice that we can read the expression left to right without having to look ahead to check for operator precedence, this is a key characteristic of postfix notation. To change precedence, we change the order of the expression, for example, to: $\boxed{2\ 3\ \times\ 1\ \mathcal{C}}$. In this case, multiplication takes priority.

Here is an example of a longer expression in both infix and postfix:

$$\boxed{2\ 4\ \mathcal{C}\ 5\ 2\ -\ -3\ \times} \equiv \boxed{\triangleright 2\ \mathcal{C}\ 4 \triangleleft - \triangleright 5 - 2 \triangleleft \times 3}\ \mathcal{D}\ 27$$

One of the chief advantages of postfix is that postfix expressions can be read left to right without having to jump ahead and look for operators of high precedence. This also means there is no need to use brackets in postfix expressions, and there is no need to designate one operator having precedence over another. What might not be obvious is how we evaluate a postfix expression using a computer. The solution to this is quite simple, we use a stack. As briefly mentioned before, a stack is a 'last in, first out' (LIFO) ordered collection of values. Two operations that are frequently defined for a stack include `pop` and `push`.

A `push` operation will add a value to the stack, and the `pop` operation removes a value from the stack, see Figure 2.4.

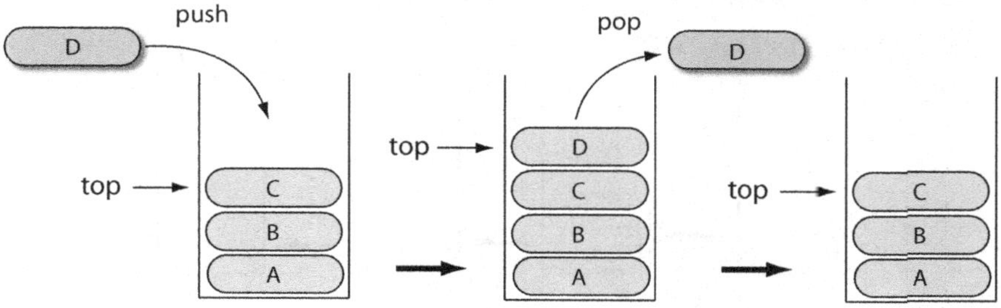

Figure 2.4 `push` and `pop` operations in a stack. `top` marks the current top of the stack

Stacks can either grow up by increasing top or grown down by decreasing top. Our stack will be designed to grow up. The pseudo-code for the pop and push operations are given below:

```
push:
   inc top
   copy element to stack

pop:
   copy element out of stack
   dec top
```

Consider the simple postfix expression $\boxed{1\ 2\ \mathcal{C}}$. To evaluate this expression using a stack, we move from left to right, and apply the following two rules:

1. Whenever we come across an operand, we push it onto the stack.
2. Whenever we come across an operator, we pop the top two operands, apply the operator to the operands and push the result on to the stack.

Figure 2.5 shows the state of the stack as we read $\boxed{1\ 2\ \mathcal{C}}$ from left to right.

Consider the expression $\boxed{1\ 4\ \mathcal{C}\ 6\ \times}$, Table 2.1 shows the order of operations are we move left to right in the expression.

How might we translate the postfix expression $\boxed{1\ 2\ \mathcal{C}}$ into bytecode? The addition operation can be assigned a single byte that will represent add, let's call this byte `add`. We will also need a push instruction to deal with the numbers 1 and 2. The question is where to keep the values 1 and 2? Since integers in the Rhodus language are 32-bit values, it would be natural to store such numbers in the 32-bit argument

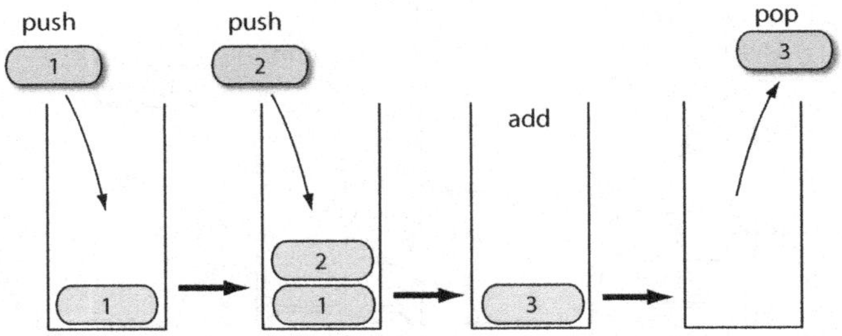

Figure 2.5 Evaluating the postfix expression 1 2 C using a stack.

Symbol	Stack	Operation
1	1	Push onto stack
4	1, 4	Push onto stack
+	5	Pop two values off the stack, compute the sum and push answer onto the stack
6	5, 6	Push onto stack
×	30	Pop two values off the stack, compute the product and push answer onto the stack

Table 2.1 Evaluating 1 4 C 6 × using a stack.

of the bytecode (Figure 2.3); and that is what we'll do. What do we do if the argument is a double or more problematic a string? Such values cannot be stored in a 32-bit space. In such cases, we will create a table of constants, and the 32-bit argument will be used to index values in the constant table. Integers will be stored directly in the 32-bit argument, and to distinguish between the 32-bit argument being used as an integer or index, we will use a specific opcode, `pushi` to mean push the 32-bit value stored in the argument onto the stack.

We can now write our first bytecode program for executing 1 2 C :

```
pushi 1
pushi 2
add
```

`pushi` will push an integer value on to the stack. For other types of constants such as doubles or strings, we'll have specific push opcodes for them, such as `pushd` and `pushs`. Figure 2.6 shows what the code will look like in memory.

For the longer expression, 1 4 C 6 × we would write the following program:

```
pushi 1
```

byte	32-bit word	byte	32-bit word	byte
`pushi`	1	`pushi`	2	`add`

Figure 2.6 Small program for executing [1 2 C] as it would appear in memory.

```
pushi 4
add
pushi 6
mult
```

At this point, we've introduced three of the opcodes that will be used by the virtual machine, these are `pushi`, `add`, and `mult`. We can complete the arithmetic operations by also including subtraction, division, power, and unary minus. Table 2.2 lists the current opcodes.

OpCode	Description
`pushi`	push an integer onto the stack.
`add`	pop two items off the stack, add them and push result onto stack.
`sub`	pop $value_1$, pop $value_2$, push ($value_2$ - $value_1$) onto the stack.
`umi`	(unary minus) pop $value_1$, push (-$value_1$) onto the stack
`mult`	pop two items off the stack, multiply them the push result onto the stack.
`divide`	pop $value_1$, pop $value_2$, push ($value_2$ / $value_1$) onto the stack.
`pow`	pop $value_1$, pop $value_2$, push ($value_2$^ $value_1$) onto the stack

Table 2.2 Initial list of opcodes for the virtual machine.

The order of values we pop off the stack is important for operators such as subtract or divide. In all cases, the **second** item popped off the stack corresponds to the **first** operand of the infix expression. For example, given the operation:

$\texttt{value}_2$ `operator` $\texttt{value}_1$

$value_1$ is the first value popped off the stack and $value_2$ is the second value popped off the stack. When we return to the source parser we'll see that it will first parse $value_2$ and output a push instruction, it will parse the operator and hold it, it will parse $value_1$ and output another push instruction, and finally output the operator. This will give us a sequence of instructions:

```
pushi value2
pushi value1
operator
```

This is the same order that the Python virtual machine uses for arithmetic operators, though other virtual machines may be different.

2.4 High-Level View of the Interpreter

We'll begin with a simple version of the final virtual machine. In later chapters, we will extend the virtual machine as new requirements emerge.

The virtual machine will support the following types:

Type	Description
Integer	Signed 32-bit
Floating Point (double)	8-byte double floating point number
Boolean	A single byte
String	A sequence of unicode characters
List	A heterogeneous list

Figure 2.7 shows a high-level view of the interpreter. There are several components in the figure that we've not yet introduced. In particular, the module and symbol table. A module is a unit that contains executable code and storage for intermediate results generated by the code. The intermediate results are stored in a symbol table that maps values to identifiers. For example, the identifier, `velocity` might be stored in the symbol table along with a float value of say `2.67`. The symbol table can store all the different data types, including integers, Booleans, floats, strings, and lists.

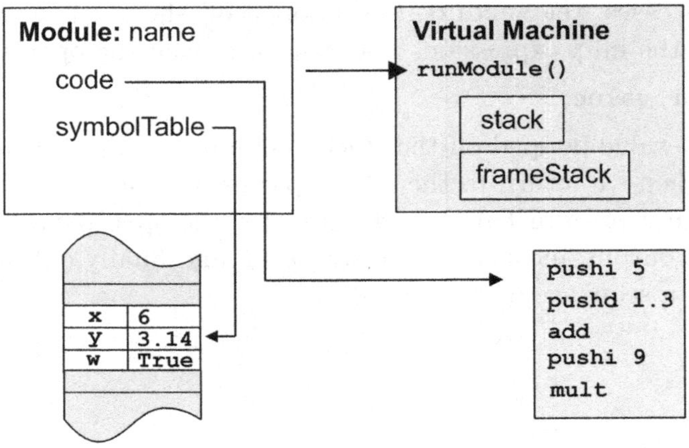

Figure 2.7 High-Level View of the Interpreter

A `module` also stores executable code. To be more exact, a module can store user-defined functions and code. For example, in Rhodus source code, a module might be written as:

```
function add (a, b)
  return a + b;
end;

x = 2.3; y = 7.8;
q = add (x, y);

println ("The sum = ", q)
```

Such source code would be converted by the interpreter into bytecode and stored in a module object. When the time comes to run the code, the module is passed to the virtual machine. The virtual machine retrieves the code and the associated symbol table and executes the code.

The virtual machine also includes several of other important components, including the runtime stack, and the frameStack that is used to support user-defined functions. In Part 3, we'll implement library supports there a program can be made from multiple modules.

2.5 An API for the Virtual Machine

That's probably enough theory for now, let's start writing some code. I often like to start by considering the API. For the virtual machine, let's create a class, `TVM`:

```
// Virtual Machine Class
TVM = class
    procedure TVM.runModule (module : TModule);
end;
```

There isn't much to the class so far, just a single procedure called `runModule` that takes a single object of type `TModule`.

2.5.1 Module

A module is a simple affair, it's just a container that holds code and an associated symbol table. A module also has a name, but in this book, the name won't be important. When the parser compiles source code, it will generate a module. At the moment, there is only a single module, but the next version of Rhodus, we will allow multiple modules that can act as libraries. The code inside a module will include user-defined functions as well as other code. The class definition for a

TModule is shown below. There isn't much to it at this stage.

```
TModule = class (TObject)
    name : string;
    code : TProgram;
    symbolTable : TSymbolTable;

    procedure   clearCode;
    constructor Create (name : string);
    destructor  Destroy; override;
end;
```

When a module is passed to runModule, the virtual machine will retrieve a reference to the code block and symbol table that the module contains. We also have a run method in the virtual machine that is called by runModule to start execution of the code. The run method is also called by the virtual machine to start the execution of user-defined functions. The code for runModule is shown below, as we can see its very simple:

```
procedure TVM.runModule (module : TModule);
begin
   self.module := module;
   run (self.module.code);
end;
```

2.5.2 TProgram

TProgram represents an array of TBytecode that we pass to the virtual machine via the module. We'll create another unit, called uCodeType.pas where we'll declare TByteCode. This is shown below:

```
unit uCodeType;

type
   TByteCode = record
         opCode : TOpCode;  // 1 byte
         index  : integer;  // 4 bytes
   end;
   TProgram = array of TByteCode;
```

We'll declare TProgram as a dynamic array of bytecodes[5] and TByteCode as a byte followed by a 32-bit integer[6].

[5] I could have used the TArray type here, but there is little to be gained using this closely related type. TArray offers additional functions such as sorting and searching which we won't need.

[6] An integer type is 32-bits long on both 32 and 64-bit platforms. If you want to make this clearer, you could declare the field using the Delphi type Int32

The opcode byte will hold the instruction code or opcode that will be of type `TOpCode` and is declared in the unit `uOpCode.pas`. The 32-bit integer will be called `index` because, more often or not, we'll actually use the argument to index other data. Integer and Boolean constants will be stored directly in the argument, but for argument types such as doubles or strings, `index` will be used to lookup values in a constant values table. `uOpCode.pas` is shown below:

```
unit uOpCodes;

interface

Uses System.SysUtils;

const
   oPushi   = byte (0);  // Push integer constant onto stack
   oAdd     = byte (1);
   oSub     = byte (2);
   oMult    = byte (3);
   oDivide  = byte (4);
   oUmi     = byte (5); // Unary minus
   oPower   = byte (6); // x^y

type
    TOpCode = byte;

implementation
end.
```

Note that the byte numbers in `uOpCodes` will change as we develop the virtual machine when we add or rearranges the instructions. Once the virtual machine is complete, the byte mapping will be finalized.

2.5.3 Machine Stack

With the bytecode defined, we next introduce the all important machine stack. The stack is the workhorse of the virtual machine and is heavily used during execution. The performance of an interpreter is greatly influenced by the performance of the machine stack, and it is important to make operations on the stack as fast as possible.

```
unit uMachineStack;

interface

Uses Classes, SysUtils;

type
```

```
TStackType = (stInteger, stDouble, stBoolean, stString);
TMachineStackRecord = record
   stackType : TStackType; // 1 byte
   case TStackType of      // Max 8 bytes
    stInteger: (iValue : integer);
    stBoolean: (bValue : boolean);
    stDouble : (dValue : double);
    stString : (sValue : TStringObject);
    stList   : (lValue : TListObject);
end;
PMachineStackRecord = ^TMachineStackRecord;
TMachineStack = array of TMachineStackRecord;
```

The above declaration defines a stack element type (`TMachineStackRecord`), and a dynamic array type, `TMachineStack`, to hold stack elements. We'll define the stack type using:

```
TMachineStack = array of TMachineStackRecord;
```

The other type declared is `PMachineStackRecord`, which is a pointer to a machine stack record: `TMachineStackRecord`.

You may be wondering why I haven't used the built-in type `TStack`? This supports all the methods we need and can hold any type we care to define. The reason has to do with performance. The stack is heavily used in the virtual machine, and we need to be able to access the stack as efficiently as possible. There is a saying in programming circles that says: premature optimization is the root of all evil, the point being that a programmer should focus on good algorithms and design first, then consider optimization. In this case, however, I know from experience that a slow stack will have a significant impact on performance. For this reason, a stack is declared as a simple array of stack elements. The built-in `TStack` uses several layers of indirection, which could affect performance. I'll be honest by admitting I've not actually tested the performance of the built-in stack type versus our homegrown one. A useful exercise might be to swap in the built-in one to see how kind of performance it has.

A stack element is capable of storing a variety of data types. In the declaration, you'll see integers, floats, Boolean, strings, and lists. Since the data values are mutually exclusive, we might as well use a variant record. A variant record allows us to use the same memory space for different variable types, this saves space and, in turn, reduces the copying burden. Without a variant, a single record will occupy 32 bytes, with the variant type space is halved to 16 bytes.

The field `stackType` specifies the current type of data held by the stack record.

Each instance of a virtual machine will have its own stack. For this reason let's add a stack variable to the `TVM` class:

```
TVM = class
   private
      stack : TMachineStack;
      stackTop : integer;
      stackSize : integer;  // Maximum depth of stack

      procedure createStack (n : integer);
      procedure freeStack;
  public
      constructor Create (_stackSize : integer);
      procedure run (code : TProgram);
end;
```

I've added a couple of other private fields to the class as well. For a start, we have the stack itself: `stack : TMachineStack`. We require two methods to create and free the stack array, namely `createStack` and `freeStack`. There is also an additional variable called `stackTop` that will hold the current value for the top of the stack. When the stack is empty, `stackTop` will have a value of -1. When we push a value on to the stack, `stackTop` will be increased by one first and then the value stored in the stack. When we pop a value off the stack, the value is retrieved, and `stackTop` is decreased by one. I've also added a variable called `stackSize`, which will hold the maximum size for the stack in the current virtual machine.

We could have packaged up the stack, including the helper routines, into a class or record. The reason why I didn't is to allow the virtual machine direct access to all the relevant fields without having to go via any indirection.

There is a range of methods we need to write to support the stack for the virtual machine. We've met two methods already, `createStack` and `freeStack`. These methods are given below:

```
procedure TVM.createStack (n : integer);
begin
  setLength (stack, n);
  stackTop := -1;
end;

procedure TVM.freeStack;
begin
  setLength (stack, 0);
end;
```

In the `createStack` method, we allocate an array to hold the stack elements. We set the top of the stack, `stackTop`, to -1, indicating an empty stack. If there is not enough memory, an out of memory exception will be issued. If that happens, we'll let the exception propagate out to the main program because this error is non-recoverable. `freeStack` does the reverse, freeing the array of stack elements.

I am sure readers may question the use of a fixed size stack, surely it would be better to have an expandable stack structure? Do we anticipate that the stack should be indefinitely expandable (up to the memory limitation of the computer itself), or should it have limits? Most arithmetic calculations will rarely use up significant amounts of stack space. The one area where stack space could be used in considerable quantities is during recursive method calls. Computing the Fibonacci sequence using the classic recursive algorithm is one example where stack space can be easily exceeded. Another instance is when variables declared in a method occupy a large amount of space. For example, static arrays are sometimes stored on the stack because once the method returns, the stack space is automatically reclaimed, which is convenient. The strategy used by many modern interpreters is to store all data on the heap, which is what CPython does, for example.

Other interpreted programming languages either implement an expandable stack such as Wren[7] or, in CPython's case, a dedicated stack is created for each user function. The size of these stacks is computed at compile time. For the initial version of Rhodus, we will use a fixed size global stack but might revisit this later on.

Out of interest, compiled languages such as Object Pascal or C/C++ use 1 MB to 8 MB as a default size for their runtime stack. For Delphi compiled programs, the default stack size is 1 MB[8].

pop and push methods The `pop` method is shown below. It checks for underflow of the stack and issues an exception if that happens. This checking will slightly slow down the pop operation, but there is little we can do to avoid it. An unchecked underflow could cause a major program crash due to illegal memory access.

The other thing to note about the `pop` method is that it returns a pointer to the stack element. This is to avoid wholesale copying of the stack record back to the caller. Sometimes the caller just needs to inspect a stack element, and copying the element would be a waste of time.

The following methods show the four different `push` methods. Each one checks for a stack overflow and issues an exception if there is one. Such checking, though unavoidable, will slightly slow the push operation. One has to remember that the pop and push operations will be called many many times during execution so that even though a single test for overflow or underflow isn't very much when called many times, the delays will add up. We'll return to the issue of performance later.

```
procedure TVM.checkStackOverflow;
begin
  if stackTop = stackSize then
     raise ERuntimeException.Create('Stack overflow error');
```

[7]https://github.com/wren-lang/wren

[8]Delphi in a Nutshell by Ray Lischner

```
function TVM.pop : PMachineStackRecord;
begin
  if stackTop > -1 then
     begin
     // Return the address of the stack element
     result := @stack[stackTop];
     dec (stackTop);
     end
  else
     raise Exception.Create('Stack underflow error');
end;
```

```
end;

procedure TVM.push (iValue : integer); // Push an integer value
begin
  inc (stackTop);
  checkStackOverflow;
  stack[stackTop].iValue := iValue;
  stack[stackTop].stackType := TStackType.stInteger;
end;
```

Here are the other **push** methods:

```
// Push a Boolean value
procedure TVM.push (bValue : boolean);
begin
  inc (stackTop);
  checkStackOverflow;
  stack[stackTop].bValue := bValue;
  stack[stackTop].stackType := TStackType.stBoolean;
end;
```

```
// push a floating point value
procedure TVM.push (dValue : double);
begin
  inc (stackTop);
  checkStackOverflow;
  stack[stackTop].dValue := dValue;
  stack[stackTop].stackType := TStackType.stDouble;
end;
```

```
// Push a string value
procedure TVM.push (sValue : TStringObject);
begin
  inc (stackTop);
  checkStackOverflow;
```

```
  stack[stackTop].sValue := sValue;
  stack[stackTop].stackType := TStackType.stString;
end;
```

The push methods are all overloaded.

2.5.4 run Function

The run function is responsible for running the bytecode. It's used by runModule and when calling a user-defined function. run takes the list of bytecodes as an argument and returns a pointer to a stack element. We'll make the run method private. There is no reason for anything outside the virtual machine to use run especially since is requires a corespondent symbol table.

```
procedure TVM.run (code : TProgram);
var ip: integer; value : PMachineStackRecord;
begin
  ip := 0;  // initialize instruction counter
  value := @noneStackType; // default return value

  while true do
    begin
    case code[ip].opCode of
       oPushi  : push (code[ip].index);
       oAdd    : addOp;
       oSub    : subOp;
       oMult   : multOp;
       oDivide : divOp;
       oUmi    : unaryminusOp;
       oPow    : pop;
       oHalt   : break;
    else
       raise Exception.Create('Unknown opcode encountered in
                  virtual machine execution loop');
    end;
    inc (ip);  // next instruction
    end;
end;
```

The run method and all other methods we've so far described will be stored in a file called uVM.pas. The run method contains a loop wrapped around a case statement. Each time the loop iterates, the instruction pointer is incremented to point to the next instruction. run works its way through the code, executing one instruction at a time. There are two initializations, one to initialize the instruction pointer to zero and a second to initialize a stack variable called value the noneStackType. The run method will always leave something on the stack when it returns, this will either be a legitimate result calculated as a result of executing the code, or if no result is available, it will push the **noneStackType** value on to the stack. The loop iterates, until it comes across a **halt** bytecode (see later).

A very important observation is that the instruction pointer, **ip**, is local to the **run** function. This is of particular importance when we call user functions because it allows us to automatically restore the instruction pointer when we return from a function. This is clearer in a later chapter when we discuss user-defined functions.

We also have a simple constructor and destructor associated with TVM, shown below:

```
constructor TVM.Create (stackSize : integer);
begin
  inherited;
  self.stackSize := stackSize;
  createStack (stackSize);
end;

destructor TVM.Destroy;
begin
  freeStack;
  inherited;
end;
```

These are responsible for setting up and destroying the machine stack.

2.6 Sample Code

Before we go further, we should manually translate some Rhodus source files into virtual machine code. This will allow us to flesh out the list of bytecodes further.

2.6.1 Assignments

First of all, consider a simple assignment statement, such as:

```
a = 5
```

There are three components to this statement, a constant 5, a variable 'a' and an assignment, '='. To deal with variables such as 'a' we will create a symbol table to store them. For constants, we will introduce another table called the constant value table.

Constant Value Table

We've already seen that if the constant is an integer, we can store it in the optional operand of the bytecode. We can do the same for Boolean variables since they are technically only one byte long. For other types of constants such as strings or floating-point values, we can't fit them into the operand space because they are too big. For these kinds of constants, we'll create a **constant value table**. Individual elements will be stored in a `TConstantValueElement` shown in the declaration below.

```
  TConstantValueType = (cpDouble, cpString);

  TConstantValueElement = class (TObject)
      valueType : TConstantValueType;
      dValue    : double;
      sValue    : TStringObject;
      lList     : TListObject;
      constructor Create (value : double); overload;
      constructor Create (value : string); overload;
      destructor Destroy; override;
  end;
```

These declarations will be stored in the file `uConstantTable.pas`. We've not yet introduced `TStringObject` or `TListObject`, but these will represent references to literal strings and lists in the constant table. Examples of literal strings and lists include:

```
  aStr = "The Matrix";
  aList = {1, True, 9,876, "Winter is Coming", {1,2,3}}
```

For floating-point, string, and list constants, we'll use the value of the bytecode operand as an index to the constant table.

We can store the constant value elements in a `TList` collection:

```
  TConstantValueTable = TList<TConstantValueElement>;

var
  constantValueTable : TConstantValueTable;
```

I'm using a `TList` to declare the constant because it has a nicer syntax for appending new elements compared to having to fiddle with `setLength` if we had used a dynamic array. Also, adding one element at a time using `setLength` is both inefficient and tends to fragment memory, whereas `TList` allocates a chunk of memory and only needs to reallocate if the length of the chunk is exceeded.

For convenience, we can create the constant table in the initialization section of `uVM.pas`. We might as well put the destructor there as well:

```
initialization
  constantValueTable := TConstantValueTable.Create;
finalization
  constantValueTable.Free;
```

```
end.
```

There is no significant runtime penalty to using a `TList` instead of a dynamic array because the table is assembled during the code generation phase, and all we're doing during runtime is accessing elements of the `TList`.

We'll need some helper routines to construct elements for the constant table, these are shown below:

```
constructor TConstantValueElement.Create (value : double);
begin
  dValue := value;
  valueType := cpDouble;
end;

constructor TConstantValueElement.Create (value : string);
begin
  sValue := TStringObject.Create (value);
  valueType := cpString;
end;
```

Symbol Table

The other component to the simple assignment `a = 5` is the variable '`a`'. Such variables, along with their values, will be stored in a symbol table, Figure 2.8.[9]

Saving Values to the Symbol Table Whenever the parser comes across a variable, it must first check to see if the variable has been previously parsed. If not, it creates a new entry in the symbol table and returns an index to indicate the variable's location in the symbol table. Since the variable '`a`' is on the left side, we will need a virtual machine instruction that can save a value to the symbol table. In addition, the result on the right-hand side must be placed on the stack ready for the save instruction to retrieve.

When the parser reads `a = 5`, it must issue two instructions, one to push the constant 5 on to the stack and a second to store the value in '`a`'. For the sake of argument, let's assume that location zero in the symbol table is used to store the value of '`a`'. A bytecode program to implement `a = 5`, would therefore be:

```
   pushi 5
   store 0
```

[9]There is one complication with respect to variables that is related to their scope (i.e their visibility). For variables used inside functions, the scope is limited to the function itself. For variables used outside functions, the scope will, for now, be assumed to be program-wide. This will be modified once we introduce the idea of modules. Let's us focus here on the former.

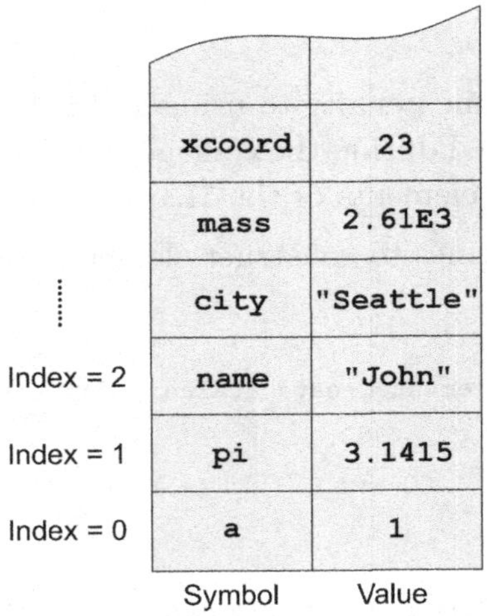

Figure 2.8 The Symbol Table.

The first instruction pushes the integer 5 on to the stack. The second instruction, `store`, will pop the value from the stack and store it in location `0` of the symbol table. Why have the `pushi` instruction, why not use an instruction such as `store 0.5`. Quite apart from the fact that we don't allow two operand instructions, the right-hand side of an assignment could be a complex expression which needs to be evaluated. Consider the expression `a = 4 + 5`. This would be turned into the following machine code:

```
pushi 4
pushi 5
add
store 0
```

Loading Values from the Symbol Table The opposite to storing is loading. What if we had an expression such as:

`a = b + 2`

Here we have a variable, `b`, on the right-hand side. To deal with this situation, we need a new bytecode instruction called `load`. Let's write down the bytecode program for this expression first:

```
load 1
pushi 2
add
store 0
```

The first instruction, `load`, takes as an argument the index to the location of the

OpCode	Description
`pushi`	push an integer onto the stack.
`pushd`	push a floating point number onto the stack
`add`	pop two items off the stack, add them and push result onto stack.
`sub`	pop $value_1$, pop $value_2$, push ($value_2$ - $value_1$) onto the stack.
`umi`	(unary minus) pop $value_1$, push (-$value_1$) onto the stack
`mult`	pop two items off the stack, multiply them the push result onto stack.
`divide`	pop $value_1$, pop $value_2$, push ($value_2$ / $value_1$) onto stack.
`pow`	pop $value_1$, pop $value_2$, push ($value_2$^ $value_1$) onto the stack
`store`	Pop the value from the stack and copy it to the symbol table. The argument represents the index to the symbol table entry.
`load`	Load the value from the symbol table and push the value onto the stack. The argument represents the index to the symbol table entry.

Table 2.3 Current list of opcodes for the virtual machine.

variable `'b'` in the symbol table. I'm assuming that `'b'` is stored in location `1` of the symbol table. `load` will retrieve the value from the symbol table and push it onto the stack. The second instruction pushes the constant 2 on to the stack followed by an add instruction which pops the two values off the stack, adds them, then pushes the result back onto the stack. Finally, `store` pops the value off the stack and stores it to the symbol table indicated by the index argument, zero.

For a final example, consider multiplying two floating-point numbers: `a = 2.3 * 5.6`.

```
pushd 0
pushd 1
mult
store 0
```

Since the arguments are floating-point numbers, these must be stored in the constant table, in this case, the value `2.3` is stored at location 0 and `5.6` at location 1 of the constant table. Note we use `pushd` to push floating-point numbers onto the stack.

Let's summarise where we are. We introduced two new instructions, `load`, and `store`, together with the concept of the symbol table and constant value table. Table 2.3 summarises the current list of machine instructions.

We'll next start looking at some of the more advanced language constructs and see how they can be translated into bytecode.

2.6.2 Conditionals

For representing conditionals and loops in bytecode, we need a series of new opcodes. Consider the conditional statement:

```
if a > 6 then
   println ("a is greater than 6")
end
```

There are several new instructions we need to introduce. The first is to deal with the print statement that will be useful for debugging purposes. We'll introduce two new opcodes, `print` and, `println`. These will be treated as special instructions because, at the moment, we don't have any notion of an external library such as an I/O library. Not only that, we don't even yet have support for built-in functions, which we could have used to define the two print statements. User-defined functions will have to wait until later. For now, we support some limited output via `print` and `println` opcodes. We'll have to add special support for the print statements in the source code parser. `print` pops an item from the stack and output it[10], `println` pops an item from the stack, outputs it followed by a newline.

The following code illustrates printing 4 and 5, followed by a newline:

```
pushi 5
print
pushi 4
println
```

The above code first pushes `5`, prints it, then pushes `4` onto the stack, prints that, then outputs a newline.

Next, we need two instructions to deal with the conditional `if a > 6 then`. The first is to test for the relation `a > 6`, and the second, if the relation test is false, we need to jump over the code in the body of the conditional statement. These two instructions are:

`isGt` – pop two items off the stack, if the second popped item is greater than the first, then push `True` onto the stack otherwise push `False`.

`jmpIfGtFalse label` – pop one item off the stack, if the item is `False` then jump forward `label` instructions.

With these new instructions, we can compile the conditional program into the following bytecode:

[10] This is not entirely accurate, the stack will also hold the number of items to output. I've omitted this stack entry in the following examples just to make the code easier to read. In the final version the stack will contain from the top, number of items to print, followed by the items. this allows us to write code like `print(a, b, c)`

```
      load 0
      pushi 6
      isGt
      jmpFalse L1
      pushs "a is greater than 6"
      println
L1: halt
```

I've also introduced two additional bytecodes, `pushs`, which pushes a string on to the stack and `halt` (which we briefly saw before), which causes the virtual machine to stop executing and return from `run`.

An important point to make is the nature of the jump itself. The jump instruction takes an argument, `L1`, in the above example. The argument is an integer stored in the operand of the bytecode. The question arises whether the value represents an absolute or relative jump? To illustrate this I've numbered each line in the code below:

```
1        load 0
2        pushi 6
3        jmpIfGtFalse L1
4        pushs "a is greater than 6"
5        println
6   L1: halt
```

If the jump value represents an absolute location, then `L1` in the argument should equal 6. If the jump label is a relative value, then the argument would equal `+3`. Since the bytecode operand is 32-bits, we can represent relative jumps within the range: `-2,147,483,648` to `2,147,483,647`. The advantage of using relative jumps is that such bytecode is relocatable. That is, we could move the instructions 1 to 6 to another location in memory, say 15 to 21, without changing the behavior of the code. If we used absolute labels, there is no way to move code without having to recompute the jump values. We will, therefore, use relative jumps for all our jump instructions.

We should also consider the conditional with the `else` clause such as:

```
if a > 6 then
   println (True)
else
   println (False)
end
```

The bytecode for an else requires an additional bytecode, the unconditional `jmp` instruction. The argument to the `jmp` instruction is a relative distance to jump. We can write the bytecode as:

```
     load 0
     pushi 6
     jmpIfGtFalse L1     # not (a > b)
     pushb True
     println
     jmp +3
L1:  pushb False   // Else clause
     println
L2:  halt
```

This code also uses the **pushb** instruction to push a Boolean value onto the stack. With these instructions defined, we can see how other language constructs would be translated into Rhodus bytecode.

2.6.3 While Loops

Consider the following while loop:

```
a = 6
while a > 2 do
  a = a - 1
end
```

We can compile the while loop into the following bytecode:

```
     pushi 6
     store 0
L1:  load 0
     pushi 2
     isGt
     jmpIfFalse L2
     load 0
     pushi 1
     sub
     store 0
     jmp   L1
L2:  halt
```

The while loop also uses the unconditional jump, `jmp`. The argument to the `jmp` instruction has a relative distance to jump of `-8` to `L1`. The `jmp` instruction allows the while loop to repeat by jumping back to the conditional test at the start of the loop. If the conditional test fails, `jmpIfFalse` will jump to just beyond the `jmp` instruction to exit the loop.

2.6.4 Repeat/Until

The repeat/until is implemented in a similar way to the while loop except that the main body of a repeat statement is executed first before carrying out the test.

```
a = 2
repeat
  a = a + 1
until a > 4
```

In the bytecode below, we compute `a = a + 1`, then test for `a > 4`. If false, the jump instruction loops back to `L1`, else it continues.

```
        pushi 2
        store 0
   L1:  load 0
        pushi 1
        add
        store 0
        load 0
        pushi 4
        isGt
        jmpIfFalse L1
        halt
```

2.6.5 For/Loops

The last looping structure to consider is the `for` loop such as the following:

```
for i = 1 to 10 do
    println (i)
end
```

The bytecode for the `for` loop is given by:

```
       pushi 1
       store 0
  L1:  load 0
       pushi 10
       isGt
       jmpIfTrue 6
       load 0
       pushi 1
       println
       inc 0
       jmp L1
       halt
```

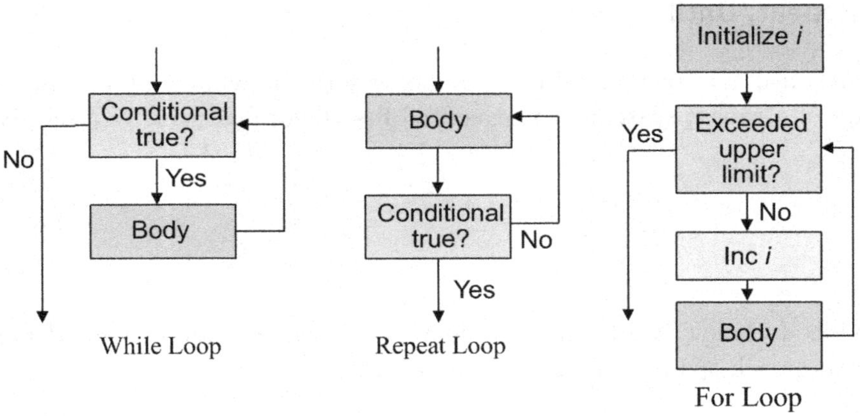

Figure 2.9 Flow charts illustrating the implementation of the while, repeat, and for loop.

To support the for loop, we need two new instructions. That first is `isLt`, which is a variant on the previously introduced `isGt`. `isLt` will pop two arguments off the stack, if the second argument is less then the first, then it pushes `True` on to the stack, otherwise, it pushes `False`. `jmpIfFalse` will pop two items off the stack, if the first item is `False`, it jumps to the label indicated by the bytecode's argument. The net effect of this is if the first stack element is less than the second stack element, then continue to the next instruction, else jump to the specified label.

The second new instruction is `inc index`. This is an optimization trick, which means we don't have to load the value on to the stack first, but the index to the variable is stored directly in the bytecode operand or in the constant. This instruction is primarily used for optimizing `for` loops.

The `for` loop is not that sophisticated at the moment but will be expanded to Rhodus version II when we include a step option.

Figure 2.9 shows the logic for each of the constructs and might help understand the byte code we used.

2.6.6 Break Statements

One thing we haven't considered in loops is break statements. For example, in the following code, the while loop is terminated when the variable '`a`' reaches 5:

```
a = 6
while a > 2 do
   a = a - 1
   if a == 5 then
      break;
```

```
end
```

A complication is that break statements can be nested and it's possible to have multiple break statement within a single loop, for example:

```
x = 12
repeat
   y = 6
   repeat
      y = y - 1
      if y == 5 then
         break;
   until y < 3;
   x = x - 1;
   if x == 7 then
      break;
until x < 6
```

When the first break is reached, control should pass to the statement immediately after the loop, in this case, `x = x - 1`. The good news is that it is possible to implement the break functionality within the current set of bytecodes, and there is no need to invoke a new special 'break' bytecode. However, all the work for handling break statements will occur at the code generator level. It shouldn't come as too much of a surprise to learn that it will involve yet another stack structure to handle nested break statements.

To implement break support, we need to be aware of two things: 1) There can be multiple break statements in a single loop; 2) Loops can be nested inside each other. Both can be dealt with using two stacks. The first stack will store the instruction pointer where a break occurs. Such a stack will be created each time we start parsing a loop structure. We'll call this stack **`breakStack`**, and it will store integers. Consider the while loop, a basic structure for parsing the while loop will look like:

```
procedure TSyntaxAnalysis.whileStatement;
var breakStack : TStack<integer>;
begin
  breakStack := TStack<integer>.Create;
  try
   ... deal with the while statement ..
  finally
    breakStack.Free;
  end;
end;
```

Each time we encounter a break statement, we emit a jump instruction with an empty placeholder for the distance we need to jump to the end of the loop. The emit instruction method, `addByteCode`, will be described in more detail in a later chapter, but it returns the instruction index where the jump instruction is located. Since there can be any number of break statements within a loop, we will be storing the location of each jump instruction that is emitted for each break statement. When we come to the end of the while loop, we need to pop each entry from the `breakStack` and patch the placeholder in the jump instruction. The following code should be placed just before the `finally` in the loop code:

```
while breakStack.Count > 0 do
    begin
    breakJump := breakStack.Pop;
    aProgram.setGotoLabel (breakJump, aProgram.getCurrentIP - breakJump);
    end;
```

There is one problem, however, which is when we have nested loops. When we come across a break statement, we need to be sure that we push the instruction pointer for the jump instruction on to the corresponding `breakStack`. If one loop is nested within another, there will be two `breakStacks`, one for the inner loop and one for the outer loop. When we push the instruction pointer onto the `breakStack`, we need to make sure we pick the right one. The way to deal with this is actually straightforward, what we need is another stack that holds the `breakStacks`. We'll call the new stack `stackOfBreakStacks`. This will be a global variable to the syntaxAnalysis class. Each time we enter a loop and create a new `breakStack`, we will push that stack onto the `stackOfBreakStacks`. When we encounter a `break` statement, the correct `breakStack` to use will be the top-most entry on `stackOfBreakStacks`. Remember that each time we exit a loop, we must pop off the `breakStack` from `stackOfBreakStacks`. With this in mind, the code for the break statement itself will be:

```
procedure TSyntaxAnalysis.breakStatement (aProgram : TProgram);
begin
  if stackOfBreakStacks.Count = 0 then
     raise ESyntaxException.Create ('Break statement illegal in this context');

  stackOfBreakStacks.Peek.push (aProgram.addByteCode (oJmp));
  sc.nextToken;
end;
```

`break` statements are only applicable within looping structures. If we encounter a break statement outside a loop, it must be an error. An easy test for this is to check whether `stackOfBreakStacks` has any entries, if not, we know that we're not inside a loop and can issue an error.

If we are inside a loop, we add a new jump instruction code. This will have a placeholder that we can patch with the distance we need to jump out of the loop.

`addByteCode` returns the current instruction pointer, and we push that value onto the top-most `breakStack` on the `stackOfBreakStacks` stack. The topmost entry will be the one we get when we use `peek`.

We need to do a few more things in the loop parsing method itself. After we create `breakStack`, we push it on to `stackOfBreakStacks`, and in the `finally` section, we pop the breakStack and free it. The code below illustrates these changes.

```
procedure TSyntaxAnalysis.whileStatement;
var breakStack : TStack<integer>;
begin
  breakStack := TStack<integer>.Create;
  stackOfBreakStacks.Push (breakStack);
  try

   ... parse the while statement ..

   expect(tEnd);
   while breakStack.Count > 0 do
       begin
       breakJump := breakStack.Pop;
       aProgram.setGotoLabel (breakJump, aProgram.getCurrentIP - breakJump);
       end;

  finally
    breakStack := stackOfBreakStacks.Pop;
    breakStack.Free;
  end;
end;
```

That completes our work on parsing and generating code for the break statement. We'll add this type of code to each loop construct. The repeat loop is very similar to the while loop, and the for loop only slightly more involved.

2.6.7 Implementing the `jmp` Instructions

Implementing the jump instructions in the virtual machine is quite straightforward. Here are the `jmp`, `jmpIsLtFalse`, and `jmpIsGtFalse` implementations:

```
oJmp : ip := ip + theProgram[ip].index - 1
oJmpIsGtFalse : begin
                item1 := pop()
                item2 := pop()
                if not (item2 > item1)
                   ip := ip + theProgram[ip].index - 1;
                end;
oJmpIsLtFalse : begin
                item1 := pop()
                item2 := pop()
```

```
        if not (item2 < item1)
           ip := ip + theProgram[ip].index - 1;
        end;
```

2.7 Initial Testing

It's time to do some testing. To keep things simple let's focus on supporting the following bytecodes in our initial version of the virtual machine:

`pushb, pushi, pushd, pushs,`
`load, store,`
`add, sub,`
`jmp, jmpIsGtFalse, jmpIsLtFalse`
and `halt`.

The `load` and `store` routines are shown below and both take the index to the symbol table entry as an argument

```
procedure TVM.store (symTabIndex : integer);
var value : PMachineStackRecord;
begin
  value := pop();
  case value.stackType of
       stInteger : storeToSymbolTable (symTabIndex, value.iValue);
       stBoolean : storeToSymbolTable (symTabIndex, value.bValue);
       stDouble  : storeToSymbolTable (symTabIndex, value.dValue);
       stString  : storeToSymbolTable (symTabIndex, value.sValue)
  else
     raise ERuntimeException.Create ('Unknown stacktype in store method');
  end;
end;
```

Note that in the `store` method we don't copy the entire stack element because `pop` returns a pointer. The code `value.iValue` is actually shorthand for `value^.iValue`; Object Pascal allows us to use this short-cut notation.

```
procedure TVM.load (symTabIndex : integer);
var symbol : PSymbol;
begin
  symbol := getFromSymbolTable (symTabIndex);
  case symbol.symbolType of
    symInteger : push (symbol.iValue);
    symDouble  : push (symbol.dValue);
    symBoolean : push (symbol.bValue);
    symString  : push (symbol.sValue);
  else
    raise ERuntimeException.Create('Unknown symbol type in load');
```

```
  end;
end;
```

The add and sub methods are fairly straightforward:

```
procedure TVM.addOp;
var st1, st2 : PMachineStackRecord;
begin
  st1 := pop();
  st2 := pop();
  case st1.stackType of
      stInteger:
         case st2.stackType of
           stInteger:
              push (st1.iValue + st2.iValue);
           stDouble:
              push (st1.iValue + st2.dValue);
         else
           raise ERuntimeException.Create ('Unrecognized type in
                         second argument in addOp');
         end;
      stDouble:
         case st2.stackType of
           stInteger:
              push (st1.dValue + st2.iValue);
           stDouble:
              push (st1.iValue + st2.dValue);
         else
             raise ERuntimeException.Create ('Unrecognized type in
                        second argument in addOp');
         end;
  else
    raise ERuntimeException.Create ('Unrecognized type in first argument in addOp );
  end;
end;
```

The only complication is that we make sure that when we sum particular types, we get an appropriate type in the result. Adding two integer results in another integer. Whereas adding an integer to a double results in a double. When we pop the two items off the stack, we check for the types and select the appropriate field from the stack element to add together. For example, if both items are integers, then we push the sum of the integer fields: `push (st1.iValue + st2.iValue)`. Later on, we'll also support the 'addition' of strings. Some combinations of types will be, however, illegal. For example, adding a Boolean to a string doesn't make sense, and we might want to report these as errors. Other languages such as CPython treat Boolean values as numerical values of 0 and 1. For example, `True + 1` in Python is a valid operation and yields the value 2.

```
procedure TVM.subOp;
var st1, st2 : PMachineStackRecord;
begin
  st1 := pop();
  st2 := pop();
  case st1.stackType of
      stInteger:
         case st2.stackType of
           stInteger:
              push (st2.iValue - st1.iValue);
           stDouble:
              push (st2.iValue - st1.dValue);
         else
            raise ERuntimeException.Create ('Unrecognized type in
                       second argument in subOp');
         end;
      stDouble:
         case st2.stackType of
           stInteger:
              push (st2.dValue - st1.iValue);
           stDouble:
              push (st2.dValue - st1.dValue);
         else
              raise ERuntimeException.Create ('Unrecognized type in
                       second argument in subOp');
         end
  else
    raise ERuntimeException.Create('Unrecognized type in first argument in subOp');
  end;
end;
```

The implementation of the subtraction bytecode is similar to the addition bytecode, except we have to be careful about the order of subtraction. The rule is that the second item popped off the stack is the first operand for the operator.

We now have the subset of bytecodes implemented. We need to create a main program to run the bytecode programs. The code for the main program is pretty simple and is given below:

```
    vm := TVM.Create;
    vm.runModule (mainModule);
    if vm.stackHasEntry then
       begin
       st := vm.pop;
       case st.stackType of
          stNone     : begin end;
          stBoolean  : if st.bValue then writeln ('True') else writeln ('False');
          stInteger  : writeln (st.ivalue);
          stDouble   : writeln (st.dValue);
          stString   : writeln (st.sValue.value);
       end;
```

```
        end;
```

We create an instance of the virtual machine and run the **runModule** method using an instance of a module that was previously passed to the parser. **runMethod** doesn't return a value, but any result we need to get access to will be expected to be on the stack of the virtual machine. I created a method **stackHasEntry** to check if there is an entry waiting to be recovered. If there is, we pop the value off then print its value.

How do we create a **program** to submit to the virtual machine? At this point in time, let's create a program using the code shown below:

```
function createProgram : TProgram;
begin
  result := [
              createByteCode (oPUSHi, 2),
              createByteCode (oPUSHi, 3),
              createByteCode (oAdd),
              createByteCode (oPop),
              createByteCode (oHalt)
            ];
end;
```

Two of the overloaded methods **createByteCode** are shown below:

```
function createByteCode (opCode : TOpCode; iValue : integer) : TByteCode;
begin
  result.opCode := opCode;
  result.index := iValue;
end;

function createByteCode (opCode : TOpCode) : TByteCode;
begin
  result.opCode := opCode;
  result.index := 0;
end;
```

There are a series of these that take into account different arguments for the bytecode. The one above deals with including an integer with the bytecode optional argument. There will be additional **createBytecode** for dealing with other types, such as Booleans, floats, and strings. For floats and strings, the method will add the constant to the constant value table and return the index that is stored with the bytecode. For example, the method for adding string constants is shown below:

```
function createByteCode (opCode : TOpCode; sValue : string) : TByteCode;
begin
  result.opCode := opCode;
  result.index := constantValueTable.Add (TStringObject.Create (sValue));
```

```
end;
```

We will return to the TStringObject type in a later chapter. If we run the simple example:

```
  pushi 2
  pushi 3
  add
  pop
```

we will get the output:

```
5
```

We could create lots of test examples like this by manually creating test programs in code, but it might be worthwhile creating a simple assembler. This is a program that converts a set of bytecode instructions in readable format into raw bytecode for the virtual machine. This will allow us to more easily write test cases as external files and integrate these into a more formal test regime.

The next chapter will be short and will focus on developing a simple assembler to help us to test the virtual machine.

3

A Simple Assembler

3.1 Assembler

Up to now we've been using an informal notation for representing bytecodes. For example, we've previously wrote the following code to implement a repeat/until loop:

```
    pushi 2
    store 0
L1: load 0
    pushi 1
    add
    store 0
    load 0
    pushi 4
    jmpIsGtFalse L1
    Halt
```

This human-readable code is called **assembler language** and the conversion of assembler into bytecode is carried out using a piece of software called an **assembler**. Assemblers are usually relatively simple pieces of software. Each line in an assembly program corresponds to a single bytecode in the virtual machine instruction code. The alternative to using assembler is to use raw bytecode. To illustrate how awkward it would be to use raw bytecode consider the simple program:

```
    pushi 12
```

```
    store 0
L1: load 0
    pushi 4
    add
    store 0
    load 0
    print
    load 0
    pushi 4
    jmpIsGtFalse L1
    Halt
```

Raw machine code, whether it be for a virtual machine or a real piece of hardware, is often expressed in hexadecimal, octal or even binary. In hexadecimal the above program is expressed as:

```
0x00 0x0000000C
0x1E 0x00000000
0x1D 0x00000000
0x00 0x00000004
0x02
0x1E 0x00000000
0x1D 0x00000000
0x00 0x00000001
0x21
0x00 0x00000004
0x19 0xFFFFFFF7
0x23
```

Not exactly easy reading. In the history of the electronic computers, assembler languages were developed very early since programming in raw machine code was simply too awkward. The early assemblers substituted a mnemonic with the equivalent machine code but over time became more sophisticated. For example, a more sophisticated assembler lets us use string labels for jump destinations instead of trying to calculate the actual jump distance by hand.

In this section, we will formalize our mnemonics and develop a simple assembler. The assembler will be useful for another reason, we can build a test framework that reads in assembly, converts it to bytecode, and passes the bytecode to the virtual machine. The assembler will only support the most basic functions that include parsing mnemonics, computing jump distances, and allow a user to specify constants such as doubles or strings.

Tables 3.1 and 3.2 shows the list of the current mnemonics for the Rhodus virtual machine. In the final virtual machine, there will be additional opcodes that deal with calling functions and supporting lists. These will be introduced in subsequent chapters.

A much more detailed description of the virtual machine mnemonics is given in the

Mnemonic	Description
Arithmetic	
`add`	pop two items off the stack, add them and push result
`sub`	pop $value_1$, pop $value_2$, push ($value_2$ - $value_1$)
`umi`	(unary minus) pop $value_1$, push (-$value_1$) onto the stack
`mult`	pop two items off the stack, multiply them and push result
`mod`	pop $value_1$, pop $value_2$ push ($value_2$ mod $value_1$)
`divide`	pop $value_1$, pop $value_2$, push ($value_2$ / $value_1$)
`pow`	pop $value_1$, pop $value_2$, push ($value_2$^ $value_1$)
Load and Store	
`store value`	pop value and copy to the symbol table. The argument represents the index to the symbol table entry.
`load value`	Load the value from the symbol table and push the value. The argument represents the index to the symbol table entry.
Logical Operations	
`and`	pop two boolean items, do logical `and`, push the result
`or`	pop two boolean items, do logical `or`, push the result
`xor`	pop two boolean items, do logical `xor`, push the result
`not`	pop one boolean item, do logical `not`, push the result
Push and pop to stack	
`pushi value`	push an integer onto the stack.
`pushb value`	push a Boolean onto the stack.
`pushd index`	push a double onto the stack, index to constant table
`pushs index`	push a string onto the stack, index to constant table
`pop`	pop item of stack
Relative Jump Operations	
`oJmp value`	Jump to new pc location
`oJmpIfTrue value`	Jump value steps if stack entry is True
`oJmpIfFalse value`	Jump value steps if stack entry is False

Table 3.1 Summary of the bytecode mnemonics. This list doesn't include byte code for lists, user and built-in functions. These will be added in later chapters.

Mnemonic	Description
Boolean Tests	
`isEq`	If `TOS-1` = `TOS` then push `True` else push `False`
`isGt`	If `TOS-1` > `TOS` then push `True` else push `False`
`isGte`	If `TOS-1` >= `TOS` then push `True` else push `False`
`isLt`	If `TOS-1` < `TOS` then push `True` else push `False`
`isLte`	If `TOS-1` <= `TOS` then push `True` else push `False`
`isNotEq`	If `TOS-1` <> `TOS` then push `True` else push `False`
Miscellaneous	
`Nop`	No operation
`Halt`	Halt the program
`print`	pop the stack and print the item to stdout
`println`	pop the stack, print the item, then output a newline to stdout

Table 3.2 Bytecode mnemonics continued. `TOS` stands for 'top of stack', so `TOS-1` represents the value of the stack one down from the top.

appendix.

As we've seen before, some of these instructions take arguments, some do not. Where opcodes expect arguments, only a single argument is expected in the form of a 32-bit integer. Assembler code can include comments, where a comment is a line that starts with a hash character, #. For example:

```
# Simple program
# Push 12 onto the stack
    pushi 12
```

The other important feature our assembler will have is the ability to specify labels. These labels are used by jump instructions. A label is a name followed by a colon, followed by the machine instruction, for example:

```
# Simple program with label
     pushi 12
 L1: print
```

The use of a colon will make it easy for us to identify labels. A jump instruction can specify a label, and the assembler will work out the correct jump distance and insert that into the generated bytecode. The astute reader may have realized that if the jump instruction occurs before a label is defined, then it is not possible for the jump distance to be computed and inserted into the bytecode. To illustrate this problem, consider the following example. In this code, we encounter the label `L1` in the jump instruction before the label is defined. This means it is not possible to compute the distance between the jmp instruction and the label.

```
# Simple program with label
      jmp L1
      ...
      ...
L1: ...
```

The solution to this problem is to design the assembler as a two-pass assembler. In the first pass, the assembler will look for labels and mark their position and name. In the second pass, the assembler will parse the instructions knowing that if it comes across a jump instruction, it can identify the label from the first pass and correctly calculate the jump distance.

Let's start writing some code. The first thing we'll do is write a function to collect information on any labels in the assembly code. We'll call this routine `collectLabels`. The idea behind this routine is that every time we come across a label of the form `name:`, we will record the name and the instruction number. For example, in the following code:

```
L1: pushi 1
    store 0
    pushi 12
L2: store 1
```

there are two labels `L1` and `L2`. `collectLabels` will record both names and their positions into a table as follows:

Name	Position
L1	0
L2	3

`collectLabels` will take a single argument, the assembler source code and is declared as:

`procedure collectLabels (const srcCode : string);`

We'll create a `TStringList` to hold the table of labels and another `TStringList` to hold the source code. The advantage of the latter is that the `TStringList` will automatically break the source code into individual lines that can be indexed by a line number. The code for setting up the `TStringList` is shown below:

```
labels := TStringList.Create;
src := TStringList.Create;
src.text := srcCode;
```

Because there could be comments in the source code, we need a separate variable, `instCounter`, to record where we are in the instruction stream. We'll set up a loop to work through the list of lines held in the `TStringList`. Inside the loop, we must first check for any empty lines or lines beginning with a hash character. In both cases, we ignore them, recall that the hash character indicates the start of a comment.

At this point, we will have a potentially valid instruction. We know that an instruction will have the syntax:

`[label:]  opcode [argument]`

where the items in square brackets are optional. This is not going to be very sophisticated, but the simplest thing to do is make sure there is only a single space between each item, after which we can split the line at the spaces. The code for doing this is:

```
  s := sl[i];
  s := RemoveDupSpaces(s);
  splitted := s.Split([' ']);
```

The `removeDupSpace` method is not a built-in Delphi method. Rather than figure one out for myself, I searched stackoverflow and found this routine which did the trick:

```
// From stackoverflow (35158485) Alex James
function RemoveDupSpaces(const Input : String) : String;
var
  P : Integer;
begin
  Result := Input;
  repeat
    P := Pos('  ', Result);  // that's two spaces
    if P > 0 then
      Delete(Result, P + 1, 1);
  until P = 0;
end;
```

The routine looks for two consecutive spaces, if it finds two spaces, it replaces it with one space. It continues this until it can't replace anyfurther.

With only single spaces, we can now use `split` to break the string into segments separated by spaces. There is the potential to find three segments corresponding to a label, an instruction, and an argument. If there is a label, it must be in the first split string, and since labels will be terminated with a colon, we can identify a label by finding a colon in the first split string. If we find a label, we strip off the colon and store the label name and the current instruction counter in the `TStringList` that holds the labels. The code for doing this is shown below:

```
if rightStr(splitted[0], 1) = ':' then
   begin
   // get label
   alabel := leftStr (splitted[0], length (splitted[0])-1);
   labels.AddObject(alabel, TObject (instCounter));
   end;
```

A trick that is probably frowned upon now but was widely used in the early days of Object Pascal is to store an integer (in this case, the instruction counter) in the object field of the string list. The alternative is to create a dedicated `TList` type that stores a record containing the name and instruction pointer. However, for such a minor application, the string list was quicker to implement. It might be worth an exercise to replace the string list with a properly engineered `TList`.

That completes the first pass for the assembler. In the second pass, we will interpret the bytecode instructions and generate the bytecode itself.

We'll create a method called `assembleCode`. This takes the source code as an argument and returns a `TProgram` type containing the raw bytecodes The first thing the method does it call `collectLabels`, which we described above. The method then loads the source code into a string list to separate the source code into individual lines. As with `collectLabels`, we ignore empty lines and comments and use split to break a line into its potentially three parts, a label, instruction, and argument which are stored in the variables, `alabel,` `opcodeStr`, and `opCodeArgument` respectively. There are different approaches we could use to identify the instructions, the one requiring the least thought is a long list of `if` statements. For example, we can check for the `add` bytecode using the line:

```
if opCodeStr = 'add' then
   begin addByteCode (oAdd, result);
   inc (instCounter);
   continue
   end;
```

We have a helper routine `addByteCode` that adds the raw bytecode information into the `TProgram` result variable, we increment the instruction counter and finally issue a `continue` that forces the code to start the next iteration of the loop.

For an instruction that has an argument, such as `pushi 12`, we use the following code:

```
if opCodeStr = 'pushi' then
   begin
   addByteCode (oPUSHi, strtoint (opCodeArgument), result);
   inc (instCounter);
   continue;
   end;
```

The only difference is the addByteCode method now also takes the optional argument. The other variant is when we encounter a jump instruction. For example, if we find a jmp instruction, the following code is called:

```
if opCodeStr = 'jmp'   then
   begin
   index := labels.IndexOf (opCodeArgument);
   if index <> -1 then
      begin
      labelInt := integer (labels.Objects[index]);
      addByteCode (oJMP, labelInt - instCounter, result);
      end
   else
      raise Exception.Create ('Unable to locate label specified in jmp opcode');
   inc (instCounter);
   continue;
   end;
```

The only complication here is the calculation of the jump distance. A jump instruction will include a label, so we must first search for the label in the labels table we constructed during the first pass. If we don't find the label, then this is an error. If we find the label, we pull out the label location from the TObject field of the labels table and then compute the difference between this value and the current instruction counter. We insert this value into the bytecode. Note that all our jump instructions are relative jumps, this is why we compute differences.

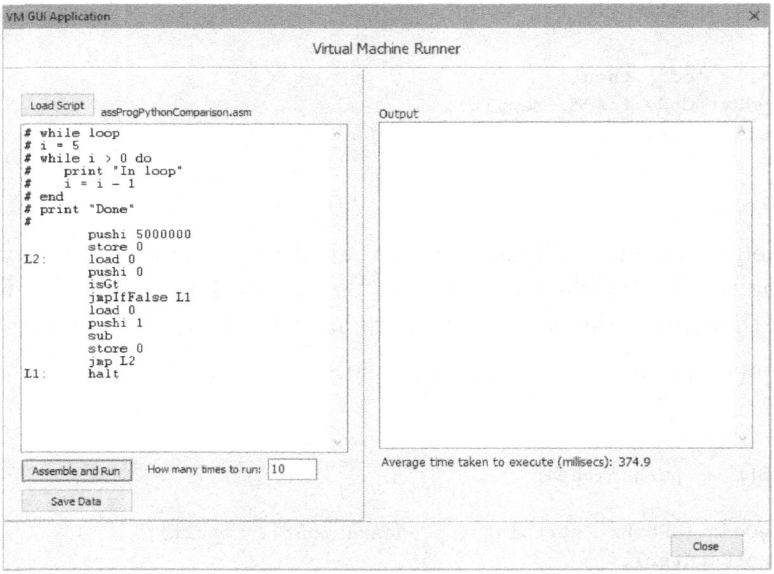

Figure 3.1 Screenshot of the virtual machine runner application.

I didn't want to spend too much time writing the assembler, so the code is a bit crude, but it is sufficient for its purpose. For example, we can't use a colon in a string argument because the assembler will confuse the colon for the end of a label. There are other circumstances where the current code won't catch badly formatted assembler code. Ideally, we should have used a lexical scanner to split up the source code in tokens, followed by a syntax parser. That would be the correct way to do it. As an exercise, it would be worth the reader rewriting the current assembler to make it much more robust and extendable.

A simple GUI was written to run assembly scripts and is shown in Figure 3.1. The GUI also makes it possible to run a script a number of times and compute the average time it took to run. In addition, the raw data can be saved to a text file. This will be useful for the next chapter when we make some detailed measurements of how fast the virtual machine is. An example of the output from executing the while script is given in Figure 3.2. This plots the raw data obtained from running a while loop 100 times. Each point represents one run of the script. Note the variability in run times. We will discuss this in the next chapter.

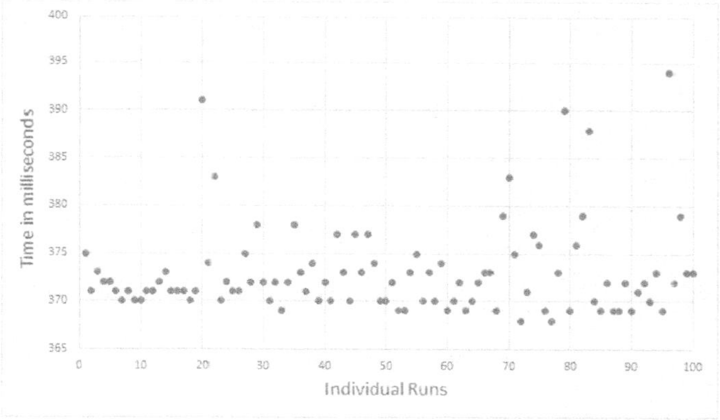

Figure 3.2 Distribution of 100 runtimes in milliseconds for a while loop.

4 Testing and Performance

4.1 Introduction

Now we come to testing. To test the code, we'll create a series of test scripts in assembler. Each test will load one of the test scripts, assemble and run the code and check for the expected result. But first, let's look at some performance considerations.

4.2 Some Performance Considerations

In general, scripting languages are slower than their compiled cousins, and design decisions at the virtual machine, and the choice of opcodes can have a profound influence on the overall performance of the interpreter. I thought I'd take this opportunity to do some performance testing and to compare, for example, how well we do with a professional interpreter such as Python.

I know from experience that the `pop` and `push` methods are called many times during the execution of a program. This means that their efficiency is important to the overall performance of the virtual machine. Both methods are quite short, which means it might be advantageous to inline these functions. The keyword **`inline`** can be appended to a method and indicates to the Delphi or FPC compiler that the code in this method may be inserted directly where the call is made. This eliminates the overhead of making the call and returning to the caller. This may make the code run faster but at the expense of a larger executable. I say 'may' because there are other factors that can affect any potential speedup. I therefore recommend that whenever you use `inline`, you test whether this actually makes any difference or not. In general, it is better to inline smaller routines that are called often.

The routines I decided to `inline` were `pop`, and the four `push` methods. For example, to inline `pop` I declared the method using the syntax:

```
function pop :  PMachineStackRecord; inline;
```

To make sure that inlining occurs, it is possible during debugging to examine the CPU window, which shows the generated code. With inlining, no call will be made to the `pop` method. Instead, the code `pop` will be found to have been inserted at the point of the call instead.

To time the performance of the inlining with and without `inline`, the `TStopWatch` class can be used that is part of the `System.Diagnostics` unit and introduced with Delphi 2010. This is a class-like record that includes a number of useful methods for timing code. To use `TStopWatch`, create an instance just as one would create an object:

```
stopwatch := TStopWatch.Create
```

However, because it's a record, there is no need to explicitly free it. `TStopWatch` gives developers access to millisecond resolution timing. There are various ways to use `TStopWatch`, but one way is shown below:

```
stopwatch := TStopWatch.Create;
stopwatch.start;                    // Start timing
.. do stuff to time ..
amountOfTime := stopwatch.ElapsedMilliseconds;
```

Note that `amountOfTime` must be of type `Int64`. We can use this code to time the `runModule` method. Does this mean we run the program once with inlining enabled and a second time with inlining absent and compare the elapsed time? The short answer is no. The first issue is that the `runModule` function runs very fast so that on short programs, the elapsed time can be of the order of a few milliseconds, if not less, which makes comparing two different situations difficult. We can solve this issue by using a more substantial machine program. A suitable program would be one that uses a loop, for example:

```
# Timing test
        pushi 1
        store 0
L1:     pushi 1
        load 0
        add
        store 0
        pushi 1
        pop
```

```
        pushi 1
        pop
        pushi 1
        pop
        pushi 1
        pop
        pushi 1
        pop
        pushi 1
        pop
        load 0
        pushi 500000
        isGt
        jmpIfFalse l1
L2:     halt
```

This code implements a while loop that loops 500,000 times. If we run the test with and without inlining of the `pop` and `push` methods, we get the two figures shown in Figure 4.1. The test was run 4000 times in order to show the variability that is possible when timing a section of code. The data shows that inlining produces a **6.5 percent** improvement in runtime. One thing you'll immediately notice about the data is that it's noisy. The first plot, in particular, shows marked deviations from the minimum values. These are generally random deviations due to the operating system running other processes and will be different each time the data is generated. Rather than take the average, which will be biased upwards by unpredictable deviations, I generally record the minimum time when comparing runs. For example, without inlining, the shortest time to compute the test took **78 ms**, whereas with inlining the test took **73 ms**. With enough samples, these numbers are reproducible. Although the difference is small, it is statistically significant. The 4000 iterations is excessive, but I wanted to show the variability in the timings. One hundred separate runs would more than suffice to get a reliable minimum time.

Inlining does not guarantee improvements in performance, which is why I made the study. For example, inlining large methods will either make no difference or in some cases, it can even reduce performance. It is important, therefore, to time the code when inlining methods.

The timings reported here were done by adjusting the code and using `TStopWatch`. There are of course, commercial and open-source specialized tools for measuring the performance of code, notably AsmProfiler, Sample Profiler, AQTime, and Nexus Quality Suite. These tools are described in more detail in Gabrijelčič's excellent book titled 'Delphi High Performance'[1].

How does this performance compare with something like Python? For this experiment I set up a while loop that looped 5,000,000 times and used the equivalent Python code shown below:

[1] https://www.packtpub.com/application-development/delphi-high-performance

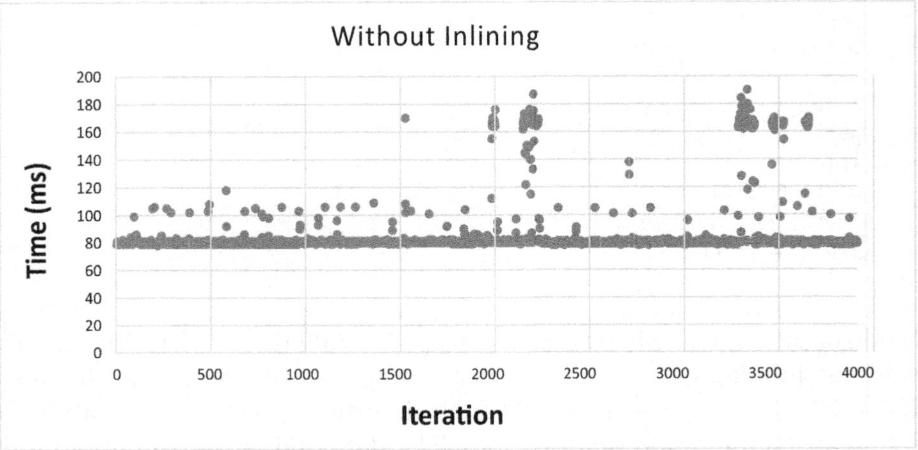

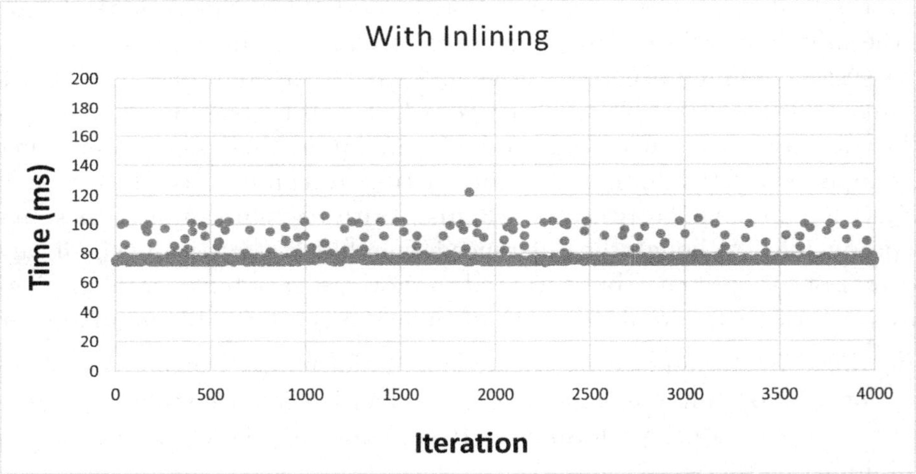

Figure 4.1 Difference in timings with and without inlining the push and pop methods. Note the large variability in timings. The data indicate that inlining results in slightly faster code than without inlining. To compare times, find the shortest time taken, do not take the average as this will bias the timings upwards due to other unpredictable processing events.

Test	Minimum Time
Rhodus	943 ms
Rhodus (no stack checking)	511 ms
Python 3.7	363 ms

Table 4.1 Timing for Rhodus and Python to execute the while loop. No stack checking means not checking for underflow and overflows when pushing an popping values.

```
# while loop
# i = 5000000
# while i > 0 do
#    print "In loop"
#    i = i - 1
# end
# print "Done"
#
        pushi 5000000 // five million
        store 0
L2:     load 0
        pushi 0
        isGt
        jmpIfFalse L1
        load 0
        pushi 1
        sub
        store 0
        jmp L2
L1:     pushs "Done"
        pushi 1
        print
        halt
```

```
import time

values = []
for i in range (100):
    t0 = time.perf_counter() # Units are seconds
    i = 1,
    while i <= 5000000:
        i = i + 1

    t1 = time.perf_counter()
    values.append ((t1-t0))
print ('Smallest time = ', min (values))
```

For our virtual machine with inlining (and as a release build), the fastest time to loop 5,000,000 times out of 50 trials was 943 ms. For the equivalent Python code, the fastest time was 362 ms (using Python 3.7 on Windows 10). This time also includes the compilation of the python code to python bytecode, but I assume this is small compared to the execution time that involves 5,000,000 iterations. This data indicates that our virtual machine is slower than Python. As an experiment, I also turned off stack overflow and underflow checking and got a much better time of 511 ms. This highlights the importance of efficient pop/push methods. All we should conclude from this study is that we're in the same ballpark in terms of performance as Python, which I think is reasonable. There are no doubt other optimizations one could do to our code, and Python over many years has been heavily optimized. To optimize further, we would need to find out where all the bottlenecks are, and to optimize, we'd focus first on the worst offenders.

4.3 Tricks for Improving Performance

Interpreters are almost always, without exception, slower than compiled code, often 2 to 50 times slower. Depending on the application, performance can be a significant issue. This has especially been the case for commercial systems such as Java or .NET where attempts have been made to make their performance almost as good as raw compiled code. In these cases, they have used just in time compilation to convert bytecode into machine code. Technologies such as LLVM can also be used effectively to generate fast interpreters. A great example is Julia, whose performance for an interpreter is extremely good. Closer to home, our own research project, libRoadRunner (`libroadrunner.org`) uses LLVM to convert differential equation models into fast executable code at runtime. We have experienced major speedups in simulation times due to this approach. However, JIT compiling is a specialized task requiring considerable expertise. If you really need performance, either invest in JIT compilation or simply use a compiler. Interpreters such as Python or R are popular not because of their speed because they are interactive, easy to use, and can be used as interfaces to C/C++ compiled code or act as glue, between compiled libraries.

There are, however, non-JIT solutions that can be used to speed up interpreters. These can be grouped into:

1. Instruction merging and specialization
2. Threading
3. Writing key parts of the interpreter loop in assembler

There is, however, always a tradeoff to increasing performance. These include increased complexity (and therefore increased maintenance costs) and potentially larger memory consumption.

Instruction merging and speciation is the easiest way to speed up an interpreter but requires a careful study of bytecode generated by the interpreter. An example of instruction merging in our own bytecode interpreter are cases such as:

```
isGt
jmpIfFalse L1
```

These two instructions could be replaced with a single instruction:

```
jumpIfGtFalse L1
```

This could be accomplished by running through the generated bytecode, looking for common patterns that could be replaced with single super-instructions. One would have to be careful to recompute the jmp distances, and a better time to do such optimization might be before actual code generation.

Another possible area to merge instructions is during arithmetic operations. For example, the following code:

```
pushi 5
pushi 7
add
```

could be replaced with:

```
pushi 5
addi 7
```

where we have introduced a new instruction **addx** that takes as an argument a constant integer value. If we relaxed the requirement that instructions can only have one optional argument, we could condense the above code even further to:

```
addy 5, 7
```

As always, performance testing should be done to check that any perceived speed increases are actually real.

Threaded Code

Our bytecode interpreter uses a large switch statement to decide what instruction to execute next. There is some overhead to using a switch statement, and a popular alternative is to use a technique called threaded code, of which there are a number of variants. The central idea is that the instruction itself contains the pointer to the method to call and also the location of the next instruction. In direct threading, when the bytecode program is loaded, each bytecode instruction is replaced with the address of the implementation of the instruction. The interpreter walks through the array of addresses, executing each one at a time. Arguments for instructions can be embedded in the array, and the array pointer adjusted accordingly. In Object Pascal we might write such as loop as:

```
pc = instructionArray
while true do
     pc ()
     inc (pc)
```

A good example of a direct threaded interpreter is the HOC interpreter found the book by Kernighan and Pike (1994) (See the section 'Useful Reading' at the end of the first chapter). There are a number variants on this theme, including indirect threading, where the bytecodes as indices into a table containing the addresses to the implementation of the instructions.

Lastly, it might be possible to replace code with raw machine assembler. In Delphi, it is possible to do this because Delphi Object Pascal allows one to write x86 assembler language as part of the Pascal code. Parts of the RTL in Delphi are written in assembler. The key thing here is to find those places where the use of assembler would make a difference. Again, testing must be carried out, particularly profiling, to find those areas that turn out to be slow. The `push` and `pop` methods would be prime candidates. The other question to consider before using raw assembly is to consider if a different approach with the existing code might not produce faster code, for example, changing the algorithm or data structures.

4.4 Testing

Forty-one basic tests for the virtual machine were put together and tested using DUnitX. These were mainly arithmetic and jmp instruction tests. These tests removed the most glaring bugs from the code. For more extensive testing, I decided to use Rhodus scripts as the main test vehicle.

To assist with testing, two new opcodes were introduced, `assertTrue` and `assertFalse`. These take an expression that must evaluate to a boolean `True` or `False`. For example:

```
assertTrue (2 + 4 == 6);
```

If the assert evaluates to True, then a `'.'` character is output to the console, otherwise an `'F'` for failure is output. Using the assert functions, tests using Rhodus scripts were based around seven groups of tests: arithmetic, boolean, conditionals, function calls, lists, loops, and strings. As of this writing, the test suite includes 219 tests. The example below shows six of the fifty arithmetic tests:

```
assertTrue (1E2 == 100);
assertTrue (1.2E2 == 120);
assertTrue (0.1E2 == 10);
assertTrue (1E-2 == 0.01);
assertTrue (1.2E-2 == 0.012);
assertTrue (-1E-2 == -0.01);
```

and the following illustrates six of the forty-two boolean tests:

```
assertTrue  (True == True);
assertTrue  (True != False);
assertFalse (True != True);
assertFalse (False != False);
assertTrue  (not False);
assertFalse (not True);
```

Of all the tests, user function and list tests were the most critical due to the complexity of these functions.

A simple REPL (Read-Eval-Print Loop) was designed to allow interactive use and testing of the interpreter. A series of commands were implemented to assist in editing and running code as well as a single command, **`tests`**, that can be invoked to run the test suite. Appendix B describes this code in more detail. For example, running all 219 tests yields the following results. Each dot represents a successful test.

```
Running tests....
Test File: conditionals.rh
..............
Test File: loops.rh
........................................
Test File: arith1.rh
..............................................
Test File: string.rh
...................
Test File: lists.rh
............................
Test File: funcCall.rh
...........................
Test File: bool.rh
.....................................
```

One problem with this test system is that if it does find a failed test, it doesn't give the line number where the test failed. This is something that could be updated in the future.

5 Emitting Code

5.1 Introduction

Up to now, we've talked about what the virtual machine looks like, as well as some of the machine instructions it can execute. We've even gone as far as to discuss what the control and loop structures in Rhodus would look like in terms of the Rhodus machine language. What we haven't touched on at all is how we're going to generate the machine instructions from Rhodus source code. For example, for an expression such as:

```
x = 1 + 2
```

the equivalent sequence of machine instructions would look like:

```
   push 2
   push 1
   add
   store x
```

The question we want to address in this chapter is how we go from Rhodus source code to the equivalent sequence of machine instructions. Let's write out a simplified parser for the previous expression. First, we need a method to parse the assignment:

```
procedure parseAssignment
begin
 if token = Identifier then
```

```
    begin
    nextToken;
    expect (equals);
    parseExpression;
    end
end;
```

Once we identify an assignment, we can pass control to another method that will parse the right-hand expression:

```
procedure parseExpression
begin
  expectNumber
  nextToken;
  while token = Plus do
    begin
    nextToken;
    expectNumber
    end;
end
```

These two methods will be able to parse code, such as:

```
a = 3
a = 3 + 4
a = 3 + 4 + 5
etc.
```

However, these two methods simply parse the source code, nothing more. There are two approaches we can now take: 1) We can emit machine instructions directly as we parse; or 2) We can encode the source code in another structure, from which we can emit the machine instructions. The most common approach is the second one, where a structure called the abstract syntax tree (AST) is constructed at compile time. It is from the AST that the machine code is generated for the virtual machine. We can expand the workflow of an interpreter, as shown in Figure 5.2.

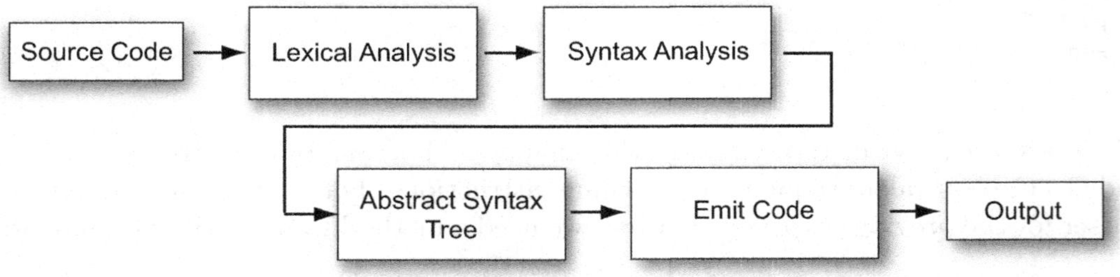

Figure 5.1 Expanded workflow for code generation.

The AST is so-called because it represents a very terse view of the source code. For example, bracketing is absent because the grouping of elements is implicit in the structure of the tree. But what is the purpose of the AST stage? There are several reasons, the first is that we get a clean separation of syntax analysis from code generation and that the AST structure can be analysed for semantic correctness even some basic code optimizations can be carried out. For languages that are statically typed, that is where variables have strict immutable types, semantic analysis can be very useful. For example, in Object Pascal, we will declare variables to be of a specific type, for example:

```
var a : integer;
```

In this case, the variable a is of type integer, and that type won't change during the lifetime of the program. This means it is possible to check for certain kinds of errors early on. In Object Pascal, we wouldn't be allowed to assign a string to the variable a. Such checks can be done at compile time so that at runtime, there are fewer things to check, which can improve the performance of the running application. Languages that have strict typing are called statically typed.

In contrast, languages such as Python are dynamically typed. A variable in Python can take on any type at runtime. The downside is that checks have to be made at runtime to ensure nothing silly is attempted. For example, adding a string to an integer is not allowed in Python at runtime. These checks can degrade performance, but the upside is that it makes it easier for users, compile-time can be a bit faster, there is no need to add declarations, making the code more succinct, and such languages are generally easier for novices to learn. In this regard, Rhodus is a dynamically typed language. For dynamically typed languages, semantic analysis is not so important though some certainly does occur.

In version one of Rhodus, we won't construct an AST, but we will generate code as we parse the source code. In Part 3, we'll change this approach. It's is still useful to understand a little bit about ASTs because it will help us appreciate better how we can generate machine instructions as we parse the source code.

5.2 Tree Structures

In computer science, a tree is a data structure that is often used to represent hierarchical information. There are many things in daily life that can be conveniently represented using a tree structure, for example, a family's genealogy, or the structure of a company.

Trees are composed of a single root and one or more branches and leaves. Unlike a real tree, the trees used in computer science are often drawn with the root at the top and the leaves at the bottom. From the root emanates two or more branches, called subtrees. At the tips of the subtrees, we can either have leaves or additional branching and further subtrees. The root, branch points, and leaves are usually represented by a node that will often contain three fields, a data field, and at least two pointer fields. If the pointer fields are `nil`, then the node is considered a leaf, otherwise, the pointers are used to point to further subtrees. Figure 5.2 shows the structure of a typical tree.

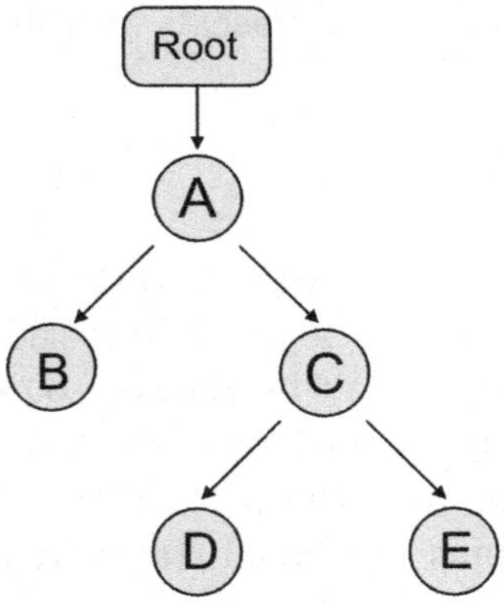

Figure 5.2 Structure of a typical tree. B, D and E are the leaves of the tree.

The simplest tree is a so-called binary tree because only two branches are allowed to emanate from a node. In Object Pascal we can represent a binary tree node using the following code:

```
PBinaryTreeNode = ^TBinaryTreeNode;
TBinaryTreeNode = record
  data : TDataField;
  leftNode  : PBinaryTreeNode;
  rigthNode : PBinaryTreeNode;
end;
```

Pascal allows us to declare a pointer to a record before the record is declared. We'll see this again in a later chapter when we discuss linked lists.

We can represent the structure of a programming language using a tree, the structure is determined by the grammar of the language. A very common illustration of this is when we represent arithmetic expressions.

Consider the simple assignment such as `a = 1 + 2`, the AST for this is shown in Figure 5.3.

The grammar rules in the Rhodus language (or any programming language) can be used to construct the abstract syntax tree. As we parse the source code, the grammar rules we used to implement the recursive decent parser, implicitly defines

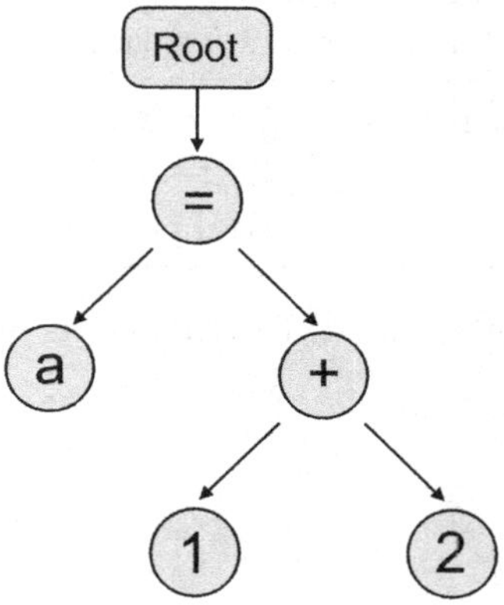

Figure 5.3 Syntax tree for `a = 1 + 2`.

the AST. That's all very well, but how would we turn an AST into executable code? Generating machine instructions from the AST involves traversing the tree in a certain order. There are three ways to 'walk' the tree, these are preorder, inorder and postorder. Consider the expression `2 + 3 * 6`. We know that in parsing this expression, the multiplication rule will be recognized first, followed by addition. In terms of a tree structure, this expression is shown in Figure 5.4.

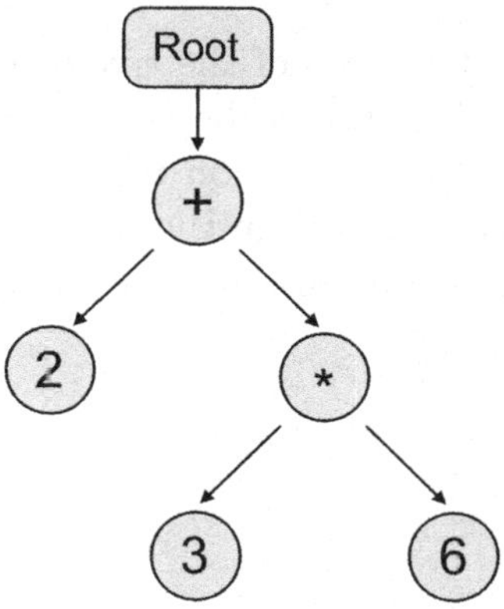

Figure 5.4 Syntax tree for `2 + 3 * 6`.

In **postorder** traversal, we first traverse the left node, followed by the right node before returning and visiting the root. Once we've traversed left, and right, we emit the symbol stored at the node. To implement this, we'd define a single procedure, called `postorder_traverse` as follows:

```
procedure postorder_traverse (node);
```

Traversal	Operations	Traversal	Operation	Traversal	Operation
Postorder	Left-subtree Right-subtree Visit Root	Preorder	Visit Root Left-subtree Right-subtree	Inorder	Left-subtree Visit Root Right-subtree

Table 5.1 Three different ways to traverse a binary tree structure.

```
begin
   if node = nil then
      exit;

   postorder_traverse (node.left);
   postorder_traverse (node.right);
   print (node.data);
end;

postorder_traverse (root);
```

You can see that the method is highly recursive, this is typical for tree algorithms. For **preorder**, we visit the root first, then traverse the left subtree, followed by the right subtree. For **inorder**, we visit the left subtree, visit the root then visit the right subtree, Table 5.1. Interestingly, inorder will recover the original infix notation.

Figure 5.5 shows the postorder traversal of the tree representing `2 + 3 * 6`. In the figure, we also show the emitted nodes as the tree is traversed. The order matches the order we'd use to evaluate the expression using a stack machine. This is how we'd generate the machine instructions from the AST. Just to be clear the postorder traversal yields the following machine instructions:

```
push 2
push 3
push 6
mult
add
```

This can be executed by the Rhodus virtual machine. Notice how the tree naturally takes care of operator precedence. This itself is a reflection of the precedence rules embedded in the Rhodus grammar.

Having done one example, we can probably see a pattern that we can use to generate code as we parse the source code without having to create the AST. Let's consider the expression `2 + 3 * 6`. The pseudo-Pascal code for parsing this is shown below:

```
procedure parseFactor
begin
```

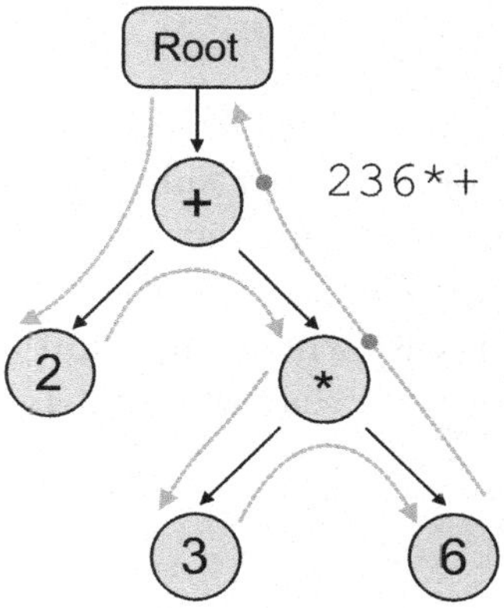

Figure 5.5 Postorder traversal of syntax tree for `2 + 3 * 6` starting at the root. Small circles mark the inner root visits.

```
  expectNumber
end

procedure parseTerm;
begin
  parseFactor
  nextToken;
  while token = mult do
    begin
    nextToken;
    parseFactor;
    end;
end;

procedure parseExpression
begin
 if token = number then
    begin
    parseTerm
    nextToken
    while token = add do
       begin
       nextToken;
       parseTerm;
       end;
    end;
end;
```

The grammar for this is just:

```
factor = number
term = factor { '*' factor }
```

```
expression = term { '+' term }
```

The rule for postorder traversal is that we process the two subtrees first, then visit the root. The subtrees could contain other subtrees. It might take a while before we return to the root we started at. This means, for example, when we parse the `expression`, we parse both terms which are in the subtrees, then emit the operator we find in the root, which in this case will be the addition operator. We do the same for terms. We can now add the code that will emit machine instructions, which is shown below.

```
procedure parseFactor
begin
  expectNumber;
  emit (push number);
end

procedure parseTerm;
begin
  parseFactor
  nextToken;
  while token = mult do
     begin
     // record the token at the root
     op := mult
     nextToken;
     parseTerm;
     emit (op);
     end;
end;

procedure parseExpression
begin
 if token = number then
    begin
    parseTerm;
    nextToken;
    while token = add do
       begin
       // record the token at the root
       op := add;
       nextToken;
       parseTerm;
       emit (op);
       end;
end;
```

Note that we only emit the root right at the end once we've traversed the two subtrees. We can construct similar code for the assignment example we looked at earlier:

```
procedure parseFactor
begin
  expectNumber;
  emit (push number);
end

procedure parseExpression
begin
  parseFactor
  nextToken;
  while token = Plus do
     begin
     op := token; // remember the token
     nextToken;
     parseFactor
     emit (op);
     end;
end

procedure parseAssignment
begin
 if token = number then
    begin
    nextToken;
    expect (equals);
    expression;
    emit (store)
    end
end;
```

5.3 While Statement

One of the advantages of this approach is that we can use it for not just arithmetic expressions but for all language constructions. Consider the Rhodus code:

```
while a < 5 do
   foo (a)
   a + a + 1
end
```

We can express this using the AST shown in Figure 5.6.

Note that nodes can have a variable number of subtrees. The call node only has one subtree, for example. We can convert this to a series of machine instructions by using postorder traversal. Applying postorder traversal to this tree yields:

```
a 5 < a foo call a 2 a + = body while
```

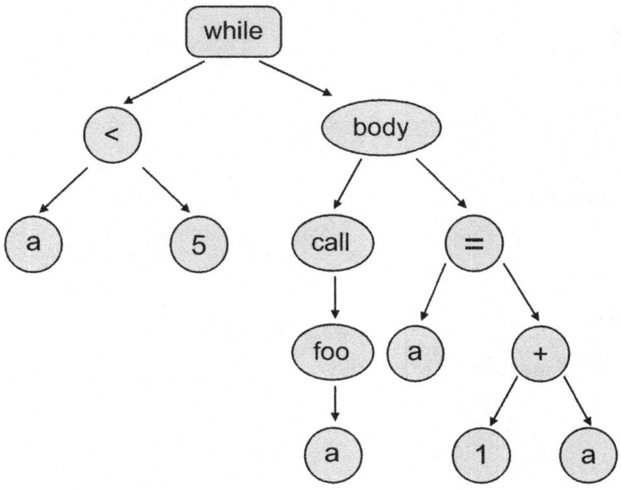

Figure 5.6 Postorder traversal of syntax tree for a while loop, function call and assignment expression.

This can be easily translated into a series of machine instructions that use a stack-based virtual machine. The rule for emitting machine instructions is the same, traverse the subtrees first, then emit the root instruction. Let's look at the code that parses and emits code for a while loop.

I've removed the logic for handling the break statements to make things easier to read. The function `whileStatement` is called when the `WHILE` token is detected. The first thing the method does is move to the next token and records the current value for the instruction pointer. At this point, it is ready to read the conditional while expression. Note that code generation of this subtree is handled elsewhere in the method expression. Once the conditional expression is processed, a `oJumpIfFalse` bytecode will be emitted to jump out of the while loop if the conditional expression is False. We then expect to see the keyword `do`, and if true, we can parse the body of the while loop, this is handled by a call to `statementList`. After that, we issue a `oJmp` bytecode and patch the relative jump back to the conditional expression at the head of the while loop. Lastly, we expect to see an `end` keyword to terminate the loop. And that's it. Not as complicated as it might seem because we pass most of the work on to other routines that handle the conditional expression and the main body of the loop. The rest of the code is making sure the syntax is correct and computing the jump instructions to make the loop work.

```
// whileStatement = WHILE expression DO statementList END
procedure TSyntaxAnalysis.whileStatement(aProgram: TProgram);
var  jumpLocation_exit, jumpLocation_back: integer;
begin
```

```
  expect (tWhile);
  again := aProgram.getCurrentIP;

  expression(aProgram);

  // record the jump location because we're going to patch the relative jump value later
  jumpLocation_exit := aProgram.addByteCode(oJmpIfFalse);

  expect(tDo);

  statementList(aProgram);

  // Record the location of the 'jump back' to jmp instruction
  JumpLocation_back := aProgram.addByteCode(oJmp);

  // Lastly, patch the relative jump instructions
  aProgram.SetGotoLabel(JumpLocation_back, again - aProgram.getCurrentIP + 1);
  aProgram.SetGotoLabel(JumpLocation_exit, aProgram.getCurrentIP - JumpLocation_exit);

  expect(tEnd);
end;
```

If we were using an AST, the code would be simpler but with significant differences. We illustrate what the code might look like below. We wouldn't be emitting any instructions, so all the calculations involving the jump instructions would be absent. Instead, the focus would be on building the `while` subtree. The big change is that all the parsing functions return a pointer to a subtree or leaf. Hence the `expression` method now returns a pointer to the subtree that represents the conditional expression. Likewise, the `statementList` method returns a pointer to the subtree representing the body of the while loop. The final thing the method does is create its own subtree from the conditional and body subtrees and combine these into a new subtree, using `makeSubTree`. This subtree is returned to whatever called `whileStatement`.

The most visible advantage of taking the AST approach is that the syntax analysis code is much similar. Of course, we relegate the missing code to a new code generation phase, but at that point, we don't have to worry about syntax anymore.

```
// whileStatement = WHILE expression DO statementList END
// Parse the while statement and construct its AST
function TSyntaxAnalysis.whileStatement(aProgram: TProgram) : PNode;
var node1, node2 : PNode;
begin
  expect (tWhile);
  node1 := expression(aProgram);
  expect(tDo);
  node2 := statementList(aProgram);
  expect(tEnd);
  result = makeSubtree (tWhile, node1, node2);
end;
```

Separating syntax analysis from code generation is recommended, but for simple languages, it is possible to combine the two and emit code as we're checking syntax. In this book, I just wanted to show that it is possible to do this, but ideally, a separation is recommended. In Part 3 of the book, we'll refactor the parser and code generation into a separate syntax analysis and code generation phase using an AST as the intermediate.

The other significant advantage of having an AST intermediate is that it is now possible to do some further analysis on the code. Whereas the syntax phase focuses on syntax compliance, the AST phase can focus on semantic issues. It is even possible to do some limited code optimization to improve performance at this stage. This will be a topic of part 3 of this book series.

5.4 Repeat Statement

Let's look at another construct, the repeat/until loop. The code for checking the syntax is quite simple and is shown below:

```
// repeatStatement = REPEAT statementList UNTIL expression
procedure TSyntaxAnalysis.repeatStatement(aProgram: TProgram);
begin
  expect (tRepeat);
  statementList (aProgram);
  expect (tUntil);
  expression (aProgram);
end;
```

As you can see, it is quite simple. For the following example:

```
  repeat
    a = a - 1;
  until a < 5;
```

The tree structure for the repeat/until loop is shown in Figure 5.7.

We can write the equivalent postorder structure based on the tree as:

```
  a a 1 + = a 5 < until repeat
```

Given this information, we can rewrite the `repeatStatement` as follows. I've omitted the logic for dealing with break statements, but it's exactly the same code as found in the while loop:

```
// repeatStatement = REPEAT statementList UNTIL expression
procedure TSyntaxAnalysis.repeatStatement(aProgram: TProgram);
var again, jumpLocation: integer;
begin
```

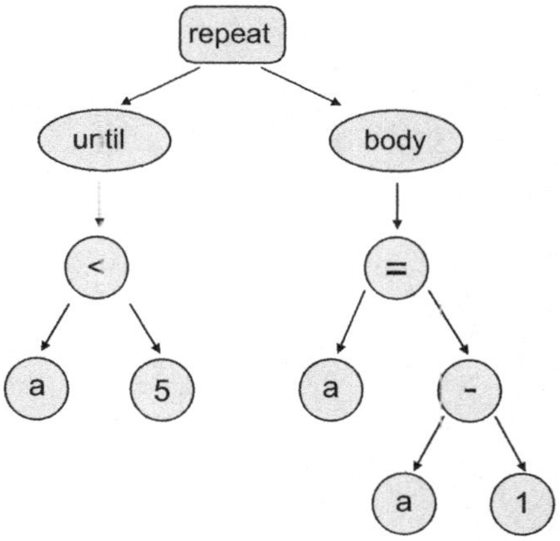

Figure 5.7 Syntax tree for the repeat/until loop.

```
  expect (tRepeat);
  again := aProgram.getCurrentInstructionPointer;
  statementList (aProgram);
  expect (tUntil);
  expression (aProgram);
  jumpLocation := aProgram.addByteCode(oJmpIfFalse);
  aProgram.setGotoLabel(jumpLocation, again - aProgram.getCurrentInstructionPointer + 1);
end;
```

As with the while loop, we delegate most of the work to other functions. The only aspect we need to deal with is setting up the jump instruction such that is the `until` statement is false, we jump back to the start of the code. Turning this into code to generate an AST is straightforward:

```
// repeatStatement = REPEAT statementList UNTIL expression
function TSyntaxAnalysis.repeatStatement(aProgram: TProgram) : PNode;
var node1, node2 : PNode;
begin
  expect (tRepeat);
  node1 := statementList (aProgram);
  expect (tUntil);
  node2 := expression (aProgram);
  result = makeSubtree (tRepeat, node1, makeSubtree (tUntil, node2, nil);
end;
```

The remaining constructs such as conditionals and `for` loops are also straightforward to code up with the code for the `for` loop being a little bit more involved.

6

User Functions

6.1 Introduction

So far we haven't talked at all about how to deal with user functions. By a user function I mean functions that look like:

```
function myfunc (x)
   return x + x
end
```

or a function with local variables:

```
function myfunc (x)
   y = 1
   return y + x
end
```

To implement functions, we only need two new machine instructions, they are:

```
   call myfunc
   ret  value
```

The first opcode calls the user function, and the second returns from the user function. So how does a function call work? Although there are only two machine instructions concerned with function calls, there is a degree of background code that we have to implement. For example, we have to consider the following four questions:

1. How do arguments get into the body of the function call?
2. What do we do about local variables?
3. How to we return values from a function?
4. How do we deal with recursion? That is `myfunc (x)` could call `myfunc (x)`.

We have to design a function call mechanism around these questions. Fortunately, all four questions can be satisfied by using the runtime stack. Not only can we pass function arguments to a function via the stack, but we can also set up storage for local variables on the stack. Let's first consider the simplest case:

```
function myfunc ()
   print ("Hello")
end
```

This example has no function arguments, no local variables, and no return value. The first thing to realize is that the code inside the function must be a separate machine program with the ability to return to the program code that called it. When we parse the source code for the Rhodus language and come across a user-defined function, we must create a separate piece of code that represents the user function. Although we're getting ahead of ourselves, we will also have to store the name of the function and details of the function in the symbol table when we do the source code parsing.

User functions will be stored in the symbol table int eh `fValue` field where the name of the function is the symbol table name. The first thing we do when making a call is to push the index that references the appropriate place in the `symbolTable` on to the stack. After that, we can initiate the call itself via the instruction `run`. Elements of the function list are records of the form:

```
TFunctionRecord = record
     name : string;
     code : TProgram;
end;
```

At the moment, this record only has the name of the user function and a reference to the user function code. When the virtual machine encounters the `call` instruction, it will call the method `callUserFunction`, which will do most of the preparatory work before handing control over to the called routine. The first version of this method is shown below:

```
procedure TVM.callUserFunction;
```

```
var funcRecord : TUserFunction;
    index : integer;
begin
  index := popInteger;

  funcRecord := module.symbolTable[index].fValue;
  run (funcRecord.code);
end;Chris Myers <myers@ece.utah.edu>
```

This function first pops off the index that was previously pushed onto the stack, using `popInteger`. `popInteger` itself will pop the value and check if it's an integer, if not, it will raise an exception as this represents a fatal error condition. It then uses index to pull out the corresponding function record and with that calls the `run` method with code as an argument. And that's it. Bear in mind this is the simplest scenario, no arguments, no local variables, and no return values. But its something we can build on.

If we assume that `myfunc` is stored at index zero in the function list, a call to `myfunc` will look like:

```
  pushi 0
  call
```

Inside `myfunc` we'll have the code:

```
  pushs "Hello"
  pushi 1      // Number of items to print
  print
  ret   none
```

The only thing that needs explanation is the `ret none` call. The return instruction must always return a value to the caller. However, if the user-function doesn't actually return anything as in this example, we instead return a none data type, called `noneStackType` in the source code.

6.2 Function Arguments

The next level of complexity to consider is how we deal with arguments in a user-defined function? For example, consider the user function:

```
function myfunc (x)
   print (2 + x)
end
```

when we call this function in source code we would write:

```
myfunc (5)
```

At the calling end, we will first push all the arguments on to the stack followed by the symbol table index and then make the call. For example:

```
pushi 5        # Arguments
pushi 0        # Index to function
call
```

The big question is, how do we reference the argument inside the user function? When we referenced module variables, we used the `load` instruction. This would take the value of the symbol from the module symbol table and push it onto the stack. We need something similar, but rather than trying to access the module symbol table, it should obtain the value from the stack since that's where the arguments will be. For this purpose, let us define a new opcode:

```
loadlocal index
```

This instruction will only be used in user functions. index will refer to the index'th argument that is stored on the stack. Since we are only pushing a single argument onto the stack in our particular example, index will be zero. With that in mind, we can write the code that will make up `myfunc` as:

```
pushi 2
locadlocal 0
add
pushi 1       // Number of items to print
print
```

One issue here is that `loadlocal` needs to know where the arguments start on the stack to index them correctly. Zero is not actually the location of the argument on the stack, the location of the argument on the stack is actually `stackTop`. What if we had more than one argument, for example:

```
y = myfunc (5, 6, 7)
```

All three arguments would be pushed onto the stack such that on entry to the user-function the stack would have the appearance:

Remember that the stack grows upwards so that the third and last argument will be at the top of the stack. `loadlocal` can be used to access any one of the three arguments. Index 0 will access `5`, index 1, `6`, and index 2 will access `7`. If we want to index the value from 0 to 2, we'll need to compute a base index. Let's call this the **base index** or `bsp`. The value for `bsp` should point to the bottom-most argument on the stack. For example, if the top of the stack before a function call was 245,

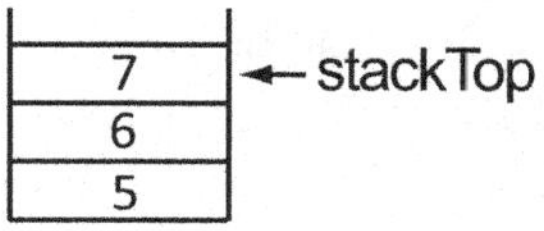

Figure 6.1 State of stack immediately after a call to `myfunc (5,6,7)`. The figure also shows the current value of `stackTop`.

then the first argument should be at location 246 on the stack. This means we should assign 246 to the base index.

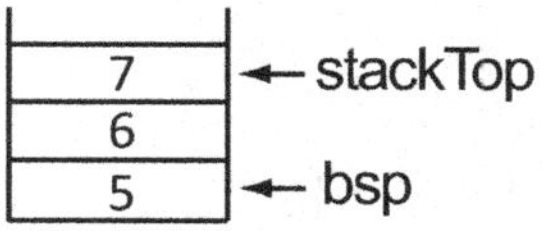

Figure 6.2 State of stack immediately after call to `myfunc (5,6,7)` with the value of `bsp` indicated. Arguments are indexed relative to `bsp`.

When we use the instruction `loadlocal 0`, what we actually mean is `locadlocal (0 + bsp)`. Before we enter the user function itself, we should compute the value of `bsp`. To do this however, we need to know the number of arguments because we'll compute `bsp` using the formula:

`bsp = stackTop - nargs + 1`

where `nargs` is the number of arguments for the user function being called. For now, Rhodus won't support a variable number of arguments in a function call, but it could if `nargs` were computed at call time rather than being computed and fixed when the source code is parsed at syntax time.

The easiest place to store the number of arguments, `nargs`, is in the user function record, so let's add a new field to that record:

```
TFunctionRecord = record
     name : string;
     nargs : integer;
     code : TProgram;
end;
```

The last question to address is where we will store the variable `bsp`? This is more important than one might realize. Let's say we store `bsp` as a private field to `TVM`. What would happen if `myfunc` calls itself or just another user function? When calling a second function, it has to go through the same sequence of operations, it did the first time it was called, including computing the value of `bsp`. There lies the problem, if we store `bsp` as a field in `TVM`, its value will be overwritten every time a user function is called. `bsp` must be preserved within a call. In other words, a user-defined function must have its own private copy of `bsp` that no other user function can touch. We could create a private `bsp` variable in `callUserFunction` but that won't do because `loadlocal` needs access to `bsp` to compute the correct index.

The solution is to create another stack that we'll call the `frameStack`,, also known as the activation record in the literature or simply the `frame`. The `frameStack` will be a place where we can safely store any kind of essential information we need for a user-function call, including `bsp`. Each time we call a user-defined function, we'll create a frame entry and push it onto the `frameStack`. Each time we exit a user-defined function we will pop it off the frame. At any particular time, the top of the `frameStack` will expose the correct frame for the currently running user-defined function.

The famestack doesn't have to be anything fancy. For now let's declare it using:

```
TFrames = record
   bsp : integer;      // base of function arguments
end;
TFrameStack = array of TFrames;
```

We'll also declare the following private variables in `TVM`:

```
frameStackTop : integer;
frameStack : TFrameStack;
```

Note that we'll need a variable to keep track of the top of the `frameStack` which we'll call `frameStackTop`. In the `TVM` constructor we'll add the two lines:

```
frameStackTop := -1;
setLength (frameStack, MAX_FRAME_DEPTH)
```

This will initialize the `frameStackTop` to empty, -1, and create some space for the frame stack itself.

In the `callUserFunction` method, we will need to compute the value for `bsp` and store it in the frame stack. This leads to the following updated `callUserFunction`:

```
procedure TVM.callUserFunction;
var funcRecord : PFunctionRecord;
    index : integer;
begin
  index := popInteger;

  funcRecord := module.symbolTable[index].fValue;

  inc (frameStackTop);
```

```
  frameStack[frameStackTop].bsp := (stackTop - funcRecord.nargs) + 1;

  run (funcRecord.code);
end;
```

The last thing to flesh out is the `loadlocal` instruction. The method that `loadLocal` calls is `localLocalSymbol`, shown below:

```
procedure TVM.localLocalSymbol (index : integer);
var bsp : integer;
begin
  bsp := frameStack[frameStackTop].bsp;
  copyToStack(@stack[index + bsp], index, @frameStack[frameStackTop]);
end;
```

It first retrieves the value of `bsp` from the current frame, then calls `copyToStack` to push the value of the argument on to the stack. What I've just said may sound a bit odd. The argument is already on the stack from when the method was called, and what we seem to be doing is pushing another copy onto the stack. The stack is doing double duty here, it acts as a memory storage place for our variables, but it's also being used as the evaluation stack. The other point to make is that when we do push the variable value on to the stack, we push a copy, this means that any change to the variable inside the user function won't change the same value outside the user function. We pass to `copyToStack` a pointer to the argument stack entry, this is done with `@stack[index + bsp])`. When we are in `copyToStack`, we check for the stack data type then make the appropriate copies. I only show two stack types in `copyToStack`(shown below), but in the final code, there will also be support for Booleans, strings, and lists. Strings and lists will take a bit more thought because they have the potential to create what's called garbage, but more of that later.

One final point is that `copyToStack` has two other arguments, `index`, and `frame`. These are used to issue appropriate error messages if we find that the value has no type. For example, if inside the function, there is a line `x = y + 1`, where `y` has not yet been assigned a value, this should be flagged as an error.

```
procedure TVM.copyToStack(stackElement: PMachineStackRecord;
                 index : integer; frame : PFrame);
begin
  inc (stackTop);
  case stackElement.stackType of
     stInteger :
         begin
         stack[stackTop].stackType := stInteger;
         stack[stackTop].iValue := stackElement.iValue;
         end;
    stDouble :
         begin
```

```
          stack[stackTop].stackType := stDouble;
          stack[stackTop].dValue := stackElement.dValue;
          end;
     stNone:
        raise ERuntimeException.Create ('Undefined variable: '
        + module.symbolTable[frame.funcIndex].fValue.symbolTable[index].symbolName);
  else
        raise Exception.Create('Internal Error: Unknown symbol in copyToStack');
  end;
end;
```

Let's look at some bytecode generated from a simple user-defined function:

```
function add (a, b, c)
   return a + b + c;
end;

println (add (1,2,3));
```

The bytecode for the user function is:

```
  0  loadLocal 0
  1  loadLocal 1
  2  add
  3  loadLocal 2
  4  add
  5  ret
```

The bytecode that calls the user-defined function is given by:

```
  0  pushi 1
  1  pushi 2
  2  pushi 3
  3  pushi 0    // Id for the user function add
  4  call
  5  pushi 1
  6  println
  7  halt
```

The mechanism we've implemented for passing the argument is called 'call by value'. What we mean by this is that a copy of the value is passed to the function so that any changes made to the value inside the function will not be reflected in changes to the same variable outside the function. The exception to this are strings and lists which are passed by reference. For example, consider the code fragment:

```
   x = 5
   myfunc (x)
```

When the call to `myfunc` is made, the value of `x` is pushed onto the stack, not a reference to `x` in the symbol table. This means that any changes we make to the value of the argument in the user function, the change should not propagate out to `x` itself. To change `x` in the current version, you'd have to return `x` from the function:

```
x = 5
x = myfunc (x)
```

List and strings will behave differently. In the following code, we can see that changes to the list inside the function is reflected by changes seen outside the function. For example:

```
function myfunc (alist)
    alist[1] = "string";
end;

x = {1,2,3};
myfunc (x);
println (x);
{1, "string", 3}
```

It's worth introducing at this point the corresponding `storelocal` opcode which takes a value from the stack and stores it to the stack:

```
storelocal index
```

This instruction does the reverse of `loadlocal`, it takes the value currently on the stack and stores it to `bsp + index`.

6.3 Local Variables

The next thing to deal with are local variables. For example, consider the user function:

```
function myfunc (x)
   y = 3
   print (x + y)
end
```

where the variable y is local to the user function `myfunc`. Since we store the function arguments on the stack, why not also store local variables on the stack? In fact, this is what we'll do. The main question is how we index the local variables? On entry to the `callUserFunction` method, we must reserve space for the local variables.

This means we need to know how many local variables there are. We'll get this information from the function record which means we have to add a new field, called `nlocals` to that function record:

```
TFunctionRecord = record
     name : string;
     nargs : integer;
     nlocals : integer;
     code : TProgram;
end;
```

`nLocals` will store the number of arguments passed to the function **plus** the number of local variables introduced in the user function. This means the actual amount of new space to reserve will be `nLocals - nArgs` since the arguments already have space reserved for them. We therefore need to increase `stackTop` by `nLocals - nArgs`. `callUserFunction` is now modified to (See Figure 6.3):

```
procedure TVM.callUserFunction;
var funcRecord : PFunctionRecord;
    index : integer;
begin
  index := popInteger;

  funcRecord := module.symbolTable[index].fValue;

  inc (frameStackTop);
  frameStack[frameStackTop].bsp := (stackTop - funcRecord.nargs) + 1;
  stackTop := stackTop + funcRecord.symbolTable.count - funcRecord.nArgs;

  run (funcRecord.code);
end;
```

Since we've lumped the arguments in with the user function local variables, we can index all the variables using a single index. One last this to do before handing control to the `run` method is to make sure that all the user function local variables are in an undefined state on entry. This is to ensure we can trap code such as:

```
     function myfunc(x)
        x = y + 1
     end;
```

where the local variable `y` has not been assigned a value even those we're using it in an expression. When we try to evaluate `y`, using the `copyToStack`, we can check to see if it's defined or not, if not, we issue an error message.

There is one problem, however. Variables outside a function are stored in the symbol table, and that symbol table stores the name of the symbol. For local variables, we don't have anywhere to keep the name of a local variable since the local variables are stored on the stack, not a symbol table. Does this matter? It does if we want to issue error messages about, for example, undefined variables, and we need to indicate which variable it was. To help with this matter, let's also define a local symbol table, its *only* purpose is to store the names of the local variables. We can keep this local symbol table in the frame structure. There is no point in creating the local symbol table at runtime as this would slow function calls down. Instead, let's create the local symbol table when we parse the source because we'll know the names of the variables at that point. During parsing, we can create and fill in the local symbol table in the function record that is kept as part of a function definition in the main symbol table. When we come to call a function, we'll copy a reference of the symbol table in the function record to a frame. The function record now contains the following fields:

```
TUserFunction = class (TObject)
     name : string;
     nArgs : integer;
     symbolTable : TSymbolTable;
     constantTable : TConstantValueTable;
     funcCode : TProgram;
end;
```

We'll describe the `constantTable` in the next section. We need to add code to `callUserFunction` to copy over the contents of the function record to the frame. The code below shows everything we'll copy from the function record to the user function frame, including the calculation of the base index, `bsp`:

```
frameStack[frameStackTop].funcIndex := index;
frameStack[frameStackTop].nArgs := funcRecord.nArgs;
frameStack[frameStackTop].nlocals := funcRecord.symbolTable.count;
frameStack[frameStackTop].constantTable := funcRecord.constantTable;
frameStack[frameStackTop].bsp := stackTop - funcRecord.nArgs + 1;
```

Remember that the local symbol table will contain the function arguments **and new** local variables.

6.4 User Function Constants

Another issue we have to worry about are constants that appear in a user-function such as Boolean, integers, doubles, and strings (Literal lists will be constructed on the fly). Outside a user-function, we store constants in a special constant table. We can do the same thing with user functions, but rather than contaminate the constant table we already have, we'll give a user function it's own constant table. As with the local symbol table, the local constant table can be constructed when

```
function myfunc (x,y)
   w = 8;
   print (x+y+w)
end;

myfunc (2,3)
```

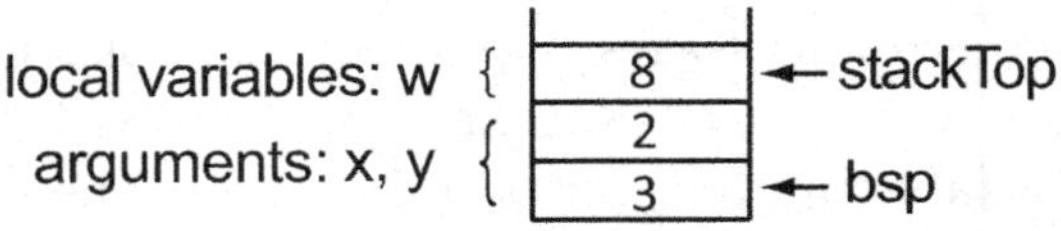

Figure 6.3 State of stack immediately after call to `myfunc (2,3)` with a single user-defined variable, w.

the source code is parsed, so there is no runtime penalty.

That completes the startup phase for calling a function. There is quite a bit of code, but most of it is just copying simple data (integer or references) and doing a few simple calculations.

6.5 Returning Values from User Functions

The next thing to consider is how do we return values from a user function. This is actually quite simple, when we encounter a return instruction, we push the value on to the stack. Alternatively, if we return from a user function without an explicit return instruction, we push a `none` value onto the stack then return. In each case, we return something. For example, let's say we have the following user function and a call to that user function shown below:

```
function myfunc()
   return 6
end;

x = myfunc()
```

This source code would generate the following bytecode:

```
call myfunc
store 0
```

Note that when the call returns, we use `store` to pop the value off the stack and store it in the symbol table entry zero. Inside the function, we'll have the following code:

```
pushi 6
ret
```

Let's deal first with the case where no value is returned. We must first return the stackTop back to where it was when the call was made. If we don't, repeated calls to user functions will eventually result in a stack overflow. We know how much to decrease the stackTop by the sum of the number of user function arguments and local variables. This means a oRet call will look like:

```
oRet : begin
       result := pop();
       // Note: nLocals include arguments and pure locals
       stackTop := stackTop - frameStack[frameStackTop].nlocals
       dec (frameStackTop);
       push(result);
       end;
```

Notice we've used frameStack to store the nargs and nlocals. This means we'll have to update the TFrameRecord and callUserFunction. When callUserFunction allocates a new frame, it needs to store nargs and nlocals into the new frame.

Once we've adjusted the value for stackTop, the final thing to do is pop the top off the frameStack by just decrementing frameStackTop. For now, that completes the task of returning from a user function. We'll revisit this when we talk more about supporting lists in the next chapter. We'll finish by listing the completed callUserFunction below:

```
procedure TVM.callUserFunction;
var
  funcRecord: TUserFunction;
  index, i: integer;
  nPureLocals : integer;
begin
  // Get the symbol table index to the function object
  index := popInteger;

  if module.symbolTable[index].symbolType = symUserFunc then
    begin
    funcRecord := module.symbolTable[index].fValue;

    inc(frameStackTop);
    if frameStackTop > MAX_FRAME_DEPTH then
       raise ERuntimeException.Create ('Exceeded maximum
                recursion depth for functions');

    // Fill in the frame entries, use with to make it more efficient
    with frameStack[frameStackTop] do
       begin
       funcIndex := index;
       nArgs := funcRecord.nArgs;
       nlocals := funcRecord symbolTable.count;
       constantTable := funcRecord.constantTable;
```

```
      bsp := stackTop - funcRecord.nArgs + 1;
      end;

    // Make space for the local variables (Note symboltable
    // contains args plus purelocals)
    stackTop := stackTop + funcRecord.symbolTable.count - funcRecord.nArgs;
    nPureLocals := funcRecord.symbolTable.count - funcRecord.nArgs;
    // Initialize the purelocals to undefined so we can detect
    // assignments to unassigned variables
    // The pure locals sit just above the arguments
    for i := 0 to nPureLocals - 1 do
        stack[frameStack[frameStackTop].bsp
                 + funcRecord.nArgs + i].stackType := stNone;

    run(funcRecord.funcCode);
    end
  else
    raise ERuntimeException.Create('Call to symbol is not a user function: ' +
      module.symbolTable[index].symbolName);
end;
```

6.6 Module Label Variables

The final thing to consider, and controversial to some, is the ability to access module-level variables from within a user function. These are in a sense, the global variables of the module. For certain sound reasons (http://heather.cs.ucdavis.edu/~matloff/globals.html), almost all languages support a mechanism that allows user functions to access variables outside the function. The key, like anything in life, is not to overdo it or abuse it. Accessing variables outside a function have its place but should only be used when no other option is available. There is further discussion of this topic on the stackoverflow [1].

In both Python and Matlab (and probably other languages), the keyword `global` is used to signify that the listed variables are actually from outside the function. We will employ the same syntax in Rhodus, for example:

```
x = 3.4; y = 6.7;

function foo()
   global y;
   y = 1.234;
   x = 9.9;
end;

println (x); // yields 3.4
```

[1] https://stackoverflow.com/questions/19158339/why-are-global-variables-evil

```
println (y); // yields 1.234
```

To declare a variable that will be accessible from outside, we use the keyword `global`, followed by a comma-separated list of variable names. The variable on the outside must exist, otherwise, it is an error. If we introduce a local variable first before declaring it as global, this is also an error. For example, the following is not allowed:

```
x = 3.4; y = 6.7;

function foo()
   y = 1.234;
   global y;
   x = 9.9;
end;
```

So how do we implement global? It's surprisingly straightforward. The first thing to note is the global variable (actually a module variable), unlike the local variety, will not be stored on the stack but will be available via the module level symbol table. When loading or storing variables in a user function, we must not use the local load and store opcodes when dealing with global variables. We will need somewhere to store the names of the global variables in order to identify them in a user function, so we'll add yet another field called `globalVariableList` to the `TUserFunction` class. The full class will now look like (I'm only showing the variable fields, not the methods:

```
   TUserFunction = class (TObject)
       name : string;
       nArgs : integer;      // should this number be on the stack?
       symbolTable : TSymbolTable;
       constantTable : TConstantValueTable;
       globalVariableList : TStringList;
       funcCode : TProgram;
end;
```

When searching `globalVariableList`, I'd like to use `find`, so we must remember to set the `sorted` property to True. This is an easy thing to forget and will result in very obscure errors if forgotten.

We deal with the global variables at the parsing level. At the code generation level, there is hardly any difference except where there is a `localLoad`, there must now be a `load`, and where there is a `localStore`, there must be a `store`.

There are two parts to the code. The first is reading the global list itself and copying the names of the global variables into the `globalVariableList` in `TUserFunction`. We also check at this stage that the global variable actually exists outside the user function.

The second part is when we parse the body of a user function and come across a variable, for example : `a = 3.4`. We need to check if the variable `a` is going to be a local variable or whether a copy is listed in `globalVariableList`. If it is listed in `globalVariableList`, we generate the global load or store opcodes. If it's actually a bonafide local variable, we generate `localLoad` or tt localStore. We have to also do this for index variables. All this decision making makes the code a bit messing and maybe later on this will be revisited.

That completes the discussion of how we'll implement user function. In the next chapter, we'll discuss how to implement string support.

7 Supporting Strings

7.1 Introduction

The first version of Rhodus will support strings. For example:

```
str1 = "Strings can be long";
str2 = "X"
```

Strings can be created and indexed. For example the following operations are supported.

```
// Create and index elements in a string:

str1 = "Merlin Engine"
print (str1[0])   // prints "M"

str1[7] = "e"
```

Note that indexing is from zero. Unlike Python and many other languages, strings in Rhodus can also be changed. For example, we can execute the following to change one of the characters in the string:

```
  a = "Robinson Crusoe"
  a[8] = "_";
  println (a)  // yields: "Robinson_Crusoe
```

There is built-in checking to stop out of range indexing, so it's not possible to change memory beyond the limits of a string. Strings can be added to other strings:

```
astr = "Space";
bstr = "is";
cstr = "big";
dstr = astr + " " + bstr + " " + cstr;
```

Strings can be multiplied by integers:

```
astr = "X";
bstr = astr*10;
print (bstr);  // will print out:  "XXXXXXXXXX"
```

An empty string is given by:

```
astr = "";
```

Strings can be passed to and from user-functions, as shown below:

```
function makeString ()
   astr = "Saturn's Rings";
   return astr
end;

blist = makeString()
```

or

```
function addStrings (astr, bstr)
   return alist + blist;
end;

print (addStrings ("Amazon", "Forest"))
```

Strings can be compared, for example:

```
astr = "1234";
bstr = "4321";
if astr == bstr then
   println ("The strings are the same")
end
```

No other functions are available to support strings in this version of Rhodus. The reason is that we have to wait for Rhodus to support libraries, at which point new functionality such as obtaining the length of a string or searching a string becomes available.

7.2 String data structure: Implementation

There are a variety of ways we could implement string support, but we're going to keep things simple. We'll first create a new type `TStringObject`:

```
TStringObject = class (TRhodusObject)

   value : string;

   function          isEqualTo (str1 : TStringObject) : boolean;
   class function    add (str1, str2 : TStringObject) : TStringObject;
   function          clone : TStringObject;
   constructor       createConstantObj (value : string);
   constructor       create (value : string);
   destructor        destroy; override;
end;
```

This is a simple class that contains just one field that holds the contents of the string. The rest of the class includes a number of methods to help us work with strings.

7.2.1 String Copy Semantics

We now come to a very important aspect about strings, memory management. Let's first describe the semantics of different situations when handling strings. Assume that the variable `a` has been assigned a string. Whenever we assign `a` to another variable, a **copy** of the string will be made.

```
a = "Robinson Crusoe"
b = a;  // A copy of the string held in a is made here
b[2] = "X"; // Change something about b, a will not be affected
```

We can now happily change the string in `b`, knowing it will not affect the string stored in `a`. The second point to make is that the original string assigned to `a` is stored in the constant table, but during the assignment to `a`, a copy is also made. Thus at the end of executing the three lines, there will be three allocated strings, one kept in the constant table, one assigned to `a`, and a third assigned to `b`.

What happens in a situation such as the following?

```
a = "Robinson Crusoe"
a = 1;
```

After we assign the string to `a`, we overwrite `a` with an integer. Remember that during the first assignment, a copy of the constant string is made, and this is what `a` is assigned to. The question is, what has happened to the memory of this copied string when we overwrite it with the integer? Or how about this situation:

```
a = "How" + "are" + "you"
```

In this code, the first two strings are concatenated to form “Howare”, after that the third string “you”, is added to form the final string. The question is, what happens to the partial string that was created when we concatenated the first two strings?

These two simple examples highlight a perennial problem when writing interpreters, and that is how to deal with allocated memory that is no longer required. Memory that seems to live in no-mans land and is no longer used is called **garbage**. The act of recovering such unused memory is called **garbage collection**.

`TStringObject` has a field called `TBlockType` that it inherited from `TRhodusOject`. This field is related to the way we're going to manage string memory.

7.3 Garbage Collection

A garbage collector is that part of an interpreter that attempts to recover previously allocated memory that is no longer in use. Whereas in programming languages such as C, or Object Pascal, we have to worry about freeing memory we no longer use, in languages such as Python or Java, there are garbage collectors that take care of freeing any unused memory. Some compiled languages such as Go also now include garbage collection, and in Object Pascal itself, strings are handled using garbage collection.

The essential idea used in garbage collection is to find allocated memory that cannot be accessed in the future. There are various strategies for doing this. The two most common are tracing and reference counting algorithms. In Rhodus, we will employ a slightly different approach, but it is worth describing briefly what the tracing and reference counting do.

In a tracing algorithm, there are designated root objects from which are traced all pathways to other objects. If certain objects cannot be traced in this way, they are deemed garbage, and their memory reclaimed. Often the roots are whatever is referenced in symbol tables. Examples of tracing algorithms include mark and sweep and tri-color marking. Ever so often, during a program run, the tracing algorithm is executed. This can sometimes cause momentary slowdowns in a program's execution as memory is being reclaimed and is one of the disadvantages of such an approach. Tracing algorithms can also be tricky to write and debug.

Another way to implement garbage collection is to use reference counting, an approach used by Python. Object Pascal uses referencing counting to handle string allocation and destruction. Reference counting is advantageous because garbage is continually being reclaimed, and there is less disruption compared to the tracing

Flag	Description
btConstant	Literal strings such as `"string"` will have their memory flag set to `btConstant`. These strings are stored in the constant table and are never freed except when the program terminates.
btBound	Any object that is bound to a symbol will have their flag set to `btBound`
btGarbage	Any object that is no longer bound to a symbol will have their flag set to `btGarbage`
btOwned	Used by lists, see next chapter

Table 7.1 Memory management flags.

algorithms. In reference counting, each object has a field called the reference count. This is a positive integer starting at zero. When an object is first created and assigned to another object, its reference count is set to one. Each time the object is referenced by a new object, its reference count increases. Each time the object loses a reference, its reference counter decreases by one. If the reference counter reaches zero, the allocated memory is reclaimed. This is a fairly straightforward algorithm to implement but does require discipline to make sure reference counts are incremented and decremented in the right places. One issue with reference counting is if two objects reference each other and no other objects reference them. In this situation, we end up with an island that can't be reclaimed.

7.3.1 Garbage Collection in Rhodus

In Rhodus, we'll take a different approach. To begin with, there will be two types of objects in Rhodus, strings and lists, and each will be managed in the same way. These objects will each inherit from `TRhodusObject`, a flag called `TBlockType`. This flag is an enumerated type containing four values:

`btGarbage, btBound, btOwned, btConstant`

The flags are described in Table 7.1.

Let's look at some situations that result in memory management needs. Let's consider these two lines of code:

```
a = "Black Hole"
a = 1;
```

The literal string "Black Hole" will be stored in the constant table and will, therefore, have its status set to `btConstant`. The `btConstant` flag ensures such strings are only freed when Rhodus fully exits.

When its time to assign a literal string to a symbol, **a copy is made**, and that copy has its status set to `btBound`. We say that the sting has been bound to the symbol. The same terminology is used when we assign a list to a variable.

When we execute the second line and assign the integer to `a`, the status of the string bound to `a` will be changed from `btBound` to `btGarbage` to indicate that the memory occupied by the string is no longer in use. What happens in the following piece of code?

```
a = "Gandalf"
b = a;
```

As before, in the first line, a copy of the literal string is made, and the copy is bound to the variable `a`. The second line assigns `a` to the variable `b`. If `b` already has an object associated with it, its status is first changed to `btGarbage`, then a copy of the string currently stored in `a` is made and assigned to `b`. The status of the copied string is set to `btBound`.

In this scenario, we end up with at least three strings, a constant string, a string assigned to `a`, a string assigned to `b`, and any garbage string that was originally bound to `b`. These operations mean that if we then execute the line:

```
b = b + " the Grey";
```

where the string in `b` is appended with the literal string "the Grey", and assigned to `b`, the original string in `b` becomes garbage. This line results in the creation of one new string and one garbage string.

Implementing these kinds of semantics is fairly straightforward. We just have to remember that when we assign a string to a variable, we make a copy and set any existing string to garbage.

One problem, what happens to all the garbage? We need to kept track of the garbage. One way to do is that each time we create a new string object, we add a reference to a list that holds the list of all allocated objects. The most efficient way to implement the list is as a linked list. Over time the linked list will get longer and longer and contain a mixture of nodes that point to constant, bound, and garbage string objects. Ever so often, we need to go through the linked list to remove and free any nodes that were set to garbage. One of the advantages of using a linked list is that removing items is efficient.

7.4 Recording Memory Allocations

Whenever a string object (or a list object) is created, it is added to a linked list. This linked list provides a record of all current memory allocations. A linked list is a convenient structure to store this information because its easy to traverse, but perhaps more importantly, it is a very efficient structure when we need to remove a single item from the list. For example, if we store the references to the allocated memory as a traditional contiguous array, if we deleted one item in the middle of the array, we'd have to move all items down one to fill in the gap. No such copying occurs with a linked link.

7.4.1 Linked Lists

As a reminder let's look at what a linked list is and how to manipulate it. Figure 7.1 shows a generic singly-linked list.

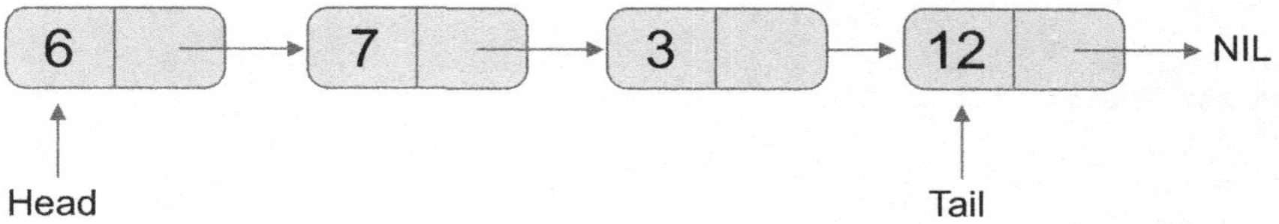

Figure 7.1 A singly linked list.

A linked list is a data structure that consists of one or more nodes. Each node is composed of two fields: a data field and a reference field. The reference field is a pointer that points to the next node in the sequence. The nodes of a linked list need not necessarily be arranged physically in sequence in memory. Instead, the sequence is dictated by the pointers as they jump from node to node. The data field can contain any kind of data and is problem specific.

A variety of linked lists have been proposed including single linked, doubly linked, and circular lists. Here, we will be only concerned with singly linked lists, as shown in Figure 7.1.

A linked list has a variable called `head` that points to the first node in the list, and sometimes a `tail` that points to the last node in the list. When there are no nodes in the list, the `head` and `tail` are set to `nil`.

In Object Pascal, the traditional way to represent a linked list is to define a record containing the data field and pointer field. This is shown below:

```
PMemoryNodePtr = TMemoryNode;
TMemoryNode = record
    robj : TRhodusObject;
```

```
    next : PMemoryNodePtr;
  end;
```

This declares a record type that represents a single node. Inside the node we have two fields, a data field, of type `TRhodusObject` and a pointer, of type `PMemoryNodePtr`. Object Pascal allows us to use a forward reference to declare the pointer to a node before we've even declared the node. Hence we declare `PMemoryNodePtr` as a `^TMemoryNode`.

Once we have a node type, we can begin to flesh out the operations we'd like to carry out. The first operation is to initialize a linked list. This is done by initializing `head` and `tail` to `nil`. Let's start a linked list class:

```
TMemoryList = class (TObject)
  private
    head, tail : PMemoryNodePtr;
  public
    constructor Create;
    destructor  Destroy; override;
end;

constructor TMemoryList.Create;
begin
  head := nil;
  tail := nil;
end;
```

We'll need a private routine to create a new node:

```
function TMemoryList.newNode (obj : TRhodusObject) : PMemoryNodePtr;
begin
  new (result);
  result^.robj := obj;
  result^.next := nil;
end;
```

This just allocates the memory, initializes the two fields, and returns a pointer to the new node. The next thing we need is the ability to add a node to the linked list. There are two situations to consider, the linked list could be empty, that is `head = nil`, or there are already nodes in the list. If the list is empty, we need to set `head` and `tail` to point to the new node. With only one node, both `head` and `tail` must point to the same node.

The second case is when we have an existing list, and we want to add a new node to the end of the list. We don't need to consider the other two possibilities, which is adding a new node to the start or middle of the list. When adding a node to the end of the list, the `tail` pointer becomes very useful. Figure 7.2 illustrates how a new node is added. We set the pointer of the last node to point to the new node, then we move the tail to the new node. The new node is now the last node.

```
function TMemoryList.addNode (robj : TRhodusObject) : PMemoryNodePtr;
var temp, p : PMemoryNodePtr;
begin
   temp := newNode(robj);
   if head = nil then  // is the linked list empty?
       begin
       head := temp;
       tail := temp;
       end
   else // Note empty then append to end of list
       begin
       tail^.next := temp;
       tail := temp;
       end;
   result := tail; // return reference to added node
end;
```

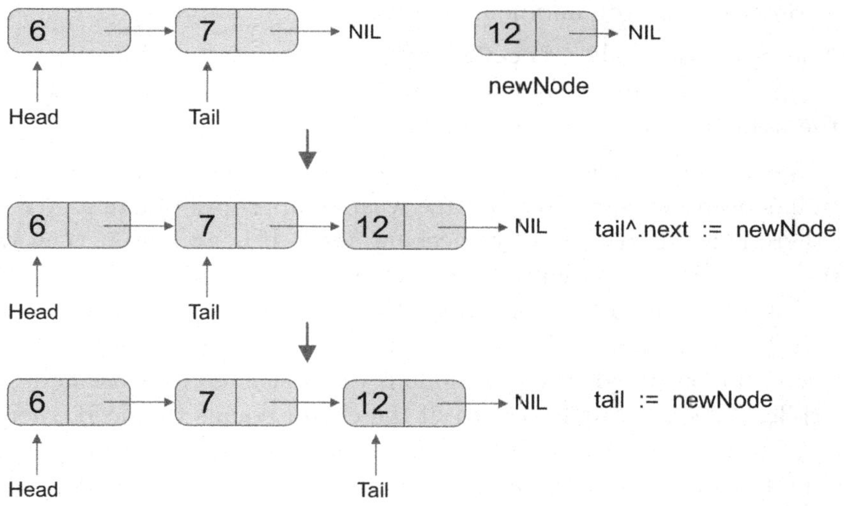

Figure 7.2 Add a new node to a linked list. Splice the new node to the end and move the tail to the new node.

A second common operation we will need is finding a particular node, such as a node indicating whether its garbage or not, and remove it from the list. To do this we must first move along the list looking for the node we wish to remove, and once found remove it. Let's first show how to move along a list. This is very straightforward and is shown in the code below:

```
procedure traverseList;
var cursor, previous : PMemoryNodePtr;
```

```
begin
   cursor := head;
   while cursor <> nil do
      begin
      previous := cursor;
      cursor := cursor^.next;
      end;
end;
```

We first assign **head** to a variable called **cursor** because we don't want to change **head** itself, **head** must remain as head. We check, as part of a while loop, if the cursor variable is **nil**, if it is we exit since it either means there is no list or we're at the end. Otherwise, we assign the variable to the node pointed to by the current node, and start the loop again.

As we traverse the list, we can inspect the nodes. This usually means checking the data field of the node. Let's say that we have a function called **ifGarbage** that returns True if the node is garbage. There are three situations to consider: the node we're looking for could be the first one in the list, the last node, or the node is buried somewhere in the middle of the list.

Let's define a function called **freeGarbage**. This function will traverse the linked list looking for nodes that have their status set to garbage. When it finds one, it deletes the node and splices the two open ends together.

freeGarbage is shown below and includes all three checks, though the logic for each operation has been relegated to a routine with an appropriate name. We first check to see if there is a list to search by checking **head**. If **head** is **nil**, then there is no list. If there is a list, we assign the current head to a new variable called **cursor**, which we'll use to track our travels along the list. So long as the cursor is not **nil**, we've nodes to check. Inside this loop, we check to see if we found a garbage node, is so we have to identify whether the node is the first node, the last node, or a node in the middle. This is straightforward, if the cursor equals the head, then we know we're at the first note. If the next pointer from the cursor is **nil**, then we know we're at the last node, everything else means we're at a middle node.

There is one other line I've added that we'll need later on which it to record what the previous node was before we jumped to the next one.

```
procedure TMemoryList.freeGarbage
var cursor, previous : PMemoryNodePtr;
begin
  if head = nil then
      exit;

  cursor := head;
  while cursor <> nil do
        begin
```

```
        if ifGarbage (cursor) then
           begin
           if head = cursor then  // We're at the first node
              freeFirstNode (cursor)
           else
              // We're in the middle or at the end
              begin
              // If we're at the end...
              if cursor^.next = nil then
                 freeLastNode (cursor, previous)
              else
                 freeMiddleNode (cursor, previous);
              end;
           end
        else
           // Node not found so continue searching
           begin
           previous := cursor;
           cursor := cursor^.next;
           end;
        end;
end;
```

Let's deal with the first and last node first; the third case might seem more tricky because when we remove a middle node, we have to splice the two open ends together.

If the node we're looking for is the first node in the list, we do two things: we move the cursor variable to point to the second node (this will be our new starting point for the list), we destroy the node we found and move **head** to point to **cursor** since this will become the new first node in the list. This is shown in the code below and illustrated in Figure 7.3:

```
procedure TMemoryList.freeFirstNode (var cursor : PMemoryNodePtr);
begin
  cursor := cursor^.next;
  head^.robj.Free;
  dispose (head);
  head := cursor;
end;
```

Notice that the `cursor` argument is a `var` argument. This is because we're updating `cursor` in the method, and that update needs to get back to the caller.

The second case is to remove the last node. To do this, we need to move the tail back one node. Recall that we recorded the previous node as we moved from one node to the next. The previous node becomes the last node, so we set its next field to `nil`. Finally, we free the node at `cursor`, which was until a moment ago, the last node. We set cursor to `nil` to indicate we're at the end of the list.

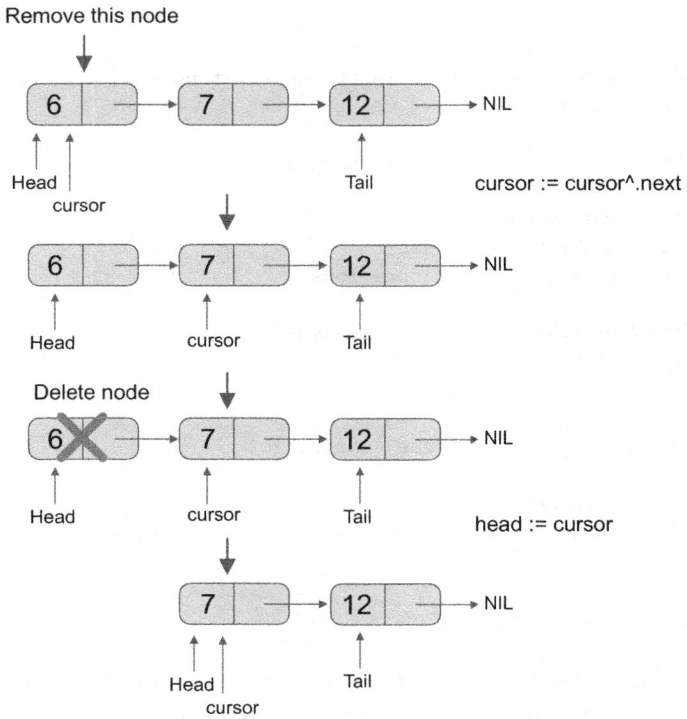

Figure 7.3 Removing the first node from a linked list.

```
procedure TMemoryList.freeLastNode  (var cursor, previous : PMemoryNodePtr);
begin
   tail := previous;
   previous^.next := nil;
   cursor^.robj.Free;
   dispose (cursor);
   cursor := nil;
end;
```

The third case involves removing a node from the middle of the list. This requires us to remove the offending node and splice the two free ends together. This is easier that one might expect. All we have to do is set the previous node pointer to the curser next node and free the node at cursor, then set cursor to the previous next node. This is illustrated in Figure 7.4.

```
procedure TMemoryList.freeMiddleNode (var cursor, previous : PMemoryNodePtr);
begin
   previous^.next := cursor^.next;
   cursor^.robj.Free;
   dispose (cursor);
```

```
  cursor := previous^.next;
end;
```

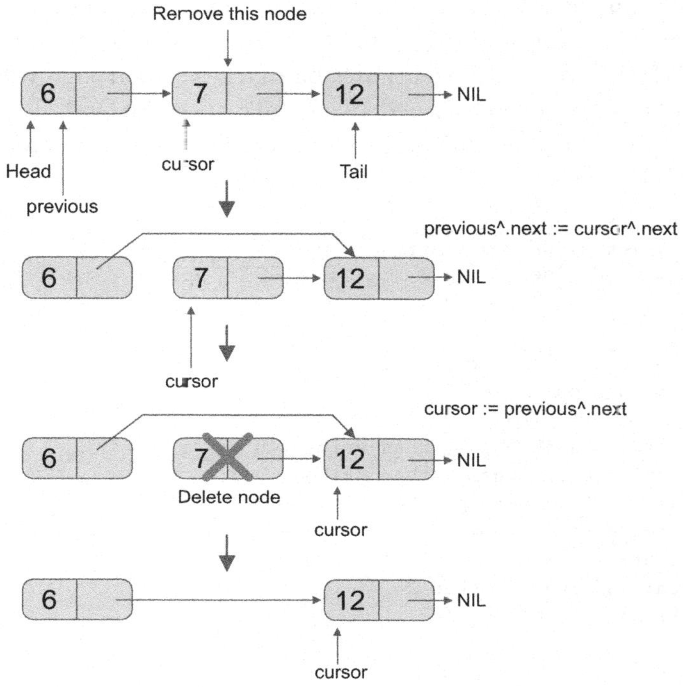

Figure 7.4 Removing a middle node from a linked list.

That completes the `freeGarbage` method. We need to consider one last operation, which is to completely free a linked list. The code for this is shown below. All we do is delete nodes one at a time as we traverse the list. We have to remember to update `head` to the next node after we delete a node. We stop when the next pointer is `nil`.

```
procedure TMemoryList.freeList;
var cursor : PMemoryNodePtr;
begin
  while head <> nil do
        begin
        cursor := head;
        head := head^.next;
        curosr^.robj.Free;
        dispose (cursor);
        end;
  head := nil; tail := nil;
end;
```

That completes all the operations we need for our linked list. Let's see how it works in practice.

Recall that each time we allocate a string or list object, we store a reference to the object in the linked list. As a script runs within the Rhodus virtual machine, memory will get allocated, and over time the linked list will get longer and longer. Many many of these nodes will eventually be marked as garbage while other nodes remain in a bound state or are owned (we'll talk about owned objects in the next chapter). To help visualize the linked list, I also wrote a simple memory map viewer, shown below:

```
function TMemoryList.mapMemory : string;
var cursor : PMemoryNodePtr;
    count : integer;
begin
  cursor := head; count := 0; result := '';
  while cursor <> nil do
        begin
        if cursor^.robj.isBound   then write ('B');
        if cursor^.robj.isOwned   then write ('O');
        if cursor^.robj.isGarbage then write ('G');
        cursor := cursor^.next;
        inc (count);
        if (count mod 70) = 0 then
           writeln;
        end;
end;
```

This returns a string of G, B, or O to indicate the status of the nodes as the list is traversed. This corresponds to garbage, bound, and owned, respectively. For example, the following script assigns a string ten times and yields the following pattern of linked memory nodes:

```
>> for i = 1 to 10 do a = "str" end
>> mem
```

GGGGGGGGGB

All nodes are garbage except for the last node that is in a bound state. This isn't surprising, each time the variable is assigned to the string, the previously assigned string becomes garbage. The final node in the linked list is in a bound state as we'd expect. For a more extensive example, the 219 test scripts were run, and the following pattern in the linked list was the result:

```
[Dynamic Memory] all:  469; garbage:  400; owned:  65; bound:  4
Memory Map:
GGGGGGGGGGGGGGGGGGGGGGGGGGGGGGGGGGGGGGGGGGGGGGGGGGGGGGGGGGGGGGGGGGGGGG
GGGGGGGGGGGGGGGGGGGGGGGGGGGGGGGGGGGGGGGGGGGGGGGGGGGGGGGGGGGGGGGGGGGGGG
```

```
GGGGGGGGGGGGGGBGGGGGGGGGGGGGGGGGGGGGGGGGGGGGGGGOGOGGGGGGGGGOGOGGOG
GGGGGGGGGOGOGOGGGOGOGOGGGOGOOGOOGOGOOGGGGOGOOGOOOGOOOGGOGOOGOOOGGGGG
GGGGGGGOGOGOOOGGGGOOOGGGGGGGGGOGGGGOGGBOGOOOOOGOOGGGGGGGOGOOGGGGOGOOGB
GBGGGGGGGGGOGOGGOGGGGGGGGGGGGGGGGGGGGGGGOGOGOGGOGGGGGGGGGGGGGGGGGGGG
GGGGGGGGGGGGGGGGGGGGGGGGGGGGGGGGGGGGGGGGGGGGGG
Memory allocated:  210204
```

There were 469 allocations in total. 400 of the allocations ended up as garbage. This may come as a surprise, but languages such as Python produce just as much, if not more garbage, as they run scripts. With computers having large memory and fast processes, we don't generally consider the amount of memory we use and throw away continually.

With longer scripts running for more time, the amount of garbage could become an issue. We, therefore, need some way to release the garbage. Since each node is marked, we can periodically search the linked list and remove any garbage nodes. The question remains, when should the garbage collector run? We will return to this question in the best chapter.

There are a couple of commands available in version 1 from the Rhodus console (see Appendix B). These are not currently part of the language but will be incorporated into a library in the next book in the series. The commands include:

Command	Description
mem	Printout a map of the currently allocated memory
free	Run the garbage collector
symbols	List all the contents of the symbol table

To give an example of the garbage collector running, let's run the collector after we've run the 219 test scripts.

```
» free
» mem
[Dynamic Memory] all:  8; garbage:  0; owned:  4; bound:  4
Memory Map:
BBOOOOBB
Memory allocated:  100648
»
```

The garbage collector has reclaimed all the reported garbage, reducing the number of persistent memory allocations to eight. However, memory was only reduced from 152,636 to 100,648. Most of the memory that remains is allocated to user functions, of which there are 25 in this case. Each user function requires a fixed overhead of 284 bytes. However, 512 instructions slots are initially allocated for the program space, and each code slot requires 5 bytes (the size of a bytecode). Therefore in total, 2560 bytes are allocated for program space per user function. This accounts for the large space occupied by the 25 user functions. This could be reduced if we removed the unused memory after we recreate a user function.

8

Supporting Lists

8.1 Introduction

The first version of Rhodus will support heterogeneous nested lists. For example:

```
alist = {1,2,3,4}
alist = {1,True, 3.14, "string"}
alist = {1, {2,3}, {4, {True, False}}}
```

Support is limited to creating lists, indexing elements, and using addition and multiplication operators. For example, the following operations are supported:

```
// Indexing elements in a list:

alist = {1,2,3,4}
print (alist[0])    // prints 1

alist[2] = True

alist = {1, {2,3}, {4, {True, False}}}
alist[1] = {4,5,6}
print (alist[2])  // prints {4, {True, False}}
```

All indexing is zero based. Any data type can be added (appended) to a list, including a list, for example:

```
alist = {1,2,3,4};
```

```
blist = alist + 5;
print (blist); // prints {1,2,3,4,5}
blist = blist + True;
print (blist); // prints {1,2,3,4,5,True}

alist = {1};
for i = 1 to 5 do
   alist = alist + i*2;
print (alist); // prints {1,2,4,6,8,10}
```

I follow the semantics used by python when adding a list to a list. This means that the elements of a list are appended to the source list. For example:

```
alist = {1,2};

blist = alist + {5,6,7};
print (blist); // prints {1,2,5,6,7}
```

If you want to append an actual list to a list, then enclose the list in another list:

```
alist = {1,2};

blist = alist + {{5,6,7}};
print (blist); // prints {1,2,{5,6,7}}
```

Lists can be multiplied by integers:

```
alist = {1,2};
blist = 3*alist;
print (blist);
// will print out:  {1, 2, 1, 2, 1, 2}
```

Note that multiplication is not like scalar multiplication of a vector. List multiplication creates multiple copies of a list. This makes it easy to create a list of a given size and content. For example:

```
alist = {2};
blist = 5*alist;
print (blist) // prints {2,2,2,2,2}
```

An empty list is given by:

```
alist = {};
blist = {} + 3;
print (blist); // prints {3}
```

Adding an empty list to another empty list yields an empty list.

Multiplying an empty list by an integer results in a list of zeros of length given by the integer. For example:

```
alist = {};
blist = 5*alist;
print (blist); // prints {0,0,0,0,0}
```

You can also add empty lists to a list:

```
alist = {};
blist = alist + {{}};
print (blist); // prints {{}}
```

Using multiplication you can generate a list of empty lists:

```
alist = {{}};
blist = 5*alist;
print (blist); // prints {{},{},{},{},{}}
```

Lists can be passed to and from user functions as shown below:

```
function makelist ()
   alist = {1,2,3,4}
   return alist
end;

blist = makelist()
```

or

```
function addlists (alist, blist)
   return alist + blist;
end;

print (addlists ({1,2,3,4}, {True, False}))
```

No other functions are available to support lists in this version of Rhodus. The reason is that we have to wait for Rhodus to support libraries at which point new functionality such as obtaining the length of a list, or search a list become available.

8.2 List data structure: Implementation

There are a variety of ways we could implement list support, but we're going to keep things simple. We'll create a new type `TListObject` that will be a list of `TListItem`. Note that `TListObject` inherits from `TRhodusObject` which holds the memory status flag (See previous chapter).

`TListItem` is a class that can store a single element of a list. To enable a list to contain lists, we have to allow `TListItem` to also store lists.

Each `TListItem` is identified using a `TListItemsType` of which there are five, `liInteger, liBoolean, liDouble, liString, liList`. Within the `TListItem` class, we also have a range of overloaded constructors for each data type.

```
TListItemType = (liInteger, liBoolean, liDouble, liString, liList);
TListItem = class;

TListObject = class (TRhodusObject)
   private
   public
     list : TObjectList<TListItem>;

     class function addLists (list1, list2 : TListObject) : TListObject;
     class function multiply (value : integer; aList : TListObject):TListObject;
     class function equals (list1, list2 : TListObject) : boolean;

     procedure   append (iValue : integer); overload;
     procedure   append (bValue : boolean); overload;
     procedure   append (dValue : double); overload;
     procedure   append (sValue : TStringObject); overload;
     function    clone : TListObject;
     function    listToString : string;
     constructor Create (count : integer);
     destructor  Destroy; override;
end;
```

8.2.1 List Copy Semantics

As with the string type described in the last chapter, lists can also create garbage requiring us to keep track of list objects. The way we deal with lists is very similar to strings except for one difference, which is that lists can include other lists.

Another difference with strings is that literal lists are not stored in the constant table but are build at runtime when executing code. This is one place where performance could be improved where we built literal lists during the syntax analysis phase. The current approach is not efficient, especially if literal lists are used inside loops. However, it wouldn't be simple to create lists at syntax time because we'd have to make sure the list is a true literal, i.e the following aren't literal: `{1,3*7}` or `{1, a}`.

Let look at some situations with respect to lists. These will be very similar to the way we handle string. Consider the code:

```
a = {1,2,3,4};
b = a;
a = 1;
```

The first line creates a list, the second line assigns a to a new variable b, and the third line overwrites a with a simple integer. In a language such as Python, the second line b = a, does not, as one might expect, copy the list to b but instead copies a reference to the list to b. In other words, a and b reference the *same* list. If we were to execute b[1] = 99, both a and b will appear to change. This is called 'copy by reference' and requires careful management of memory. When a is assigned to the integer, does it mean that the memory for the list must be freed? No, because b is still referring to the list. This requires the Python runtime system to keep track of who is referring to what memory objects so that it doesn't inadvertently free memory it shouldn't. Python solves this problem by using a reference count on each object. When we assign a = {1,2,3,4}, the reference count for the list goes up by one, when we do the second assignment, b = a the reference count goes up by one again. In the third line, the reference count for the list goes down by one. If the reference count reaches zero, then Python can safely free the memory occupied by the list object. Rhodus does something different, for an assignment such as b = a, a copy of a is made. We still have to keep track of what objects are bound to variables, and what objects are free. To show this behavior in operation, consider the code:

```
  a = {1,2,3,4};
  b = a;
  a[1] = 89;
  println (a, b)
{1,89,3,4}  {1,2,3,4}
```

In this code, only a has changed, showing that b has its own copy. As with strings, list objects inherit from TRhodusObject so that lists also have associated with them the memory status flag, which can be one of four values: btBound, btGarbage, btOwned, and btConstant. For the current implementation of lists, btConstant doesn't apply because literal lists are created at runtime. Consider the code:

```
a = {1,2,3,4};
```

When a literal list is first created, it has a memory status of btGarbage. When assignment occurs to the variable a, if the list has the garbage flag set, then we assign it to the variable and set the status of the list to btBound. If we now assign a to another variable b, this time, the list has the bound flag set so that we make a copy of the list stored in a and assign it to the variable b. The new copy of the list also acquires the btBound status.

The rules for assignment are, therefore:

Before Assignment	After Assignment
btGarbage	Assign and set to btBound
btBound	Make copy, assign and set to btBound

A small complication occurs when dealing with nested lists, particulary when we extract a sublist from a larger list. Consider the following case:

```
  a = {1,2,{3,4}};
  b = a[2];
  a[2] = 89;
  println (a, b)
{1,2,89} {3,4}
```

To handle this situation we make a copy of the sublist {3,4} and set its memory status to btBound once it's assigned to b.

Another more fundamental difference with strings is nesting. One list or string can be nested inside another list. For nested objects, the memory block status is set to btOwned. If the top-level list is set to garbage, then any garbage collection that occurs will delete the top-level and change the status of all the owned to garbage so that the next time garbage collection occurs, the previously owned elements are also deleted. This is done this way because we can't guarantee that the owned nodes aren't stored before the current cursor location in the linked list. If we did try to free them, we'd end up with node on the link list with bad pointers. The way to full free an unassigned nested list is to run the garbage collector twice.

8.2.2 Supporting Methods

In TListObject, we also have a method called listToString, which we use to print a list in text form. This method is recursive in nature in order to deal with nested lists. This method is shown below:

```
function TListObject.listToString  : string;
var i : integer;
begin
  result := '{';
  if self.Count > 0 then
     result := result + self[0].elementToString;

  for i := 1 to self.Count  - 1 do
      result := result + ',' + self[i].elementToString;
  result := result + '}';
end;
```

The above code also shows a style I use to often generate comma separated lists. I start by constructing the first element of the list, after which I begin a `for` loop. Since I've already dealt with the first element, the `for` loop starts with the second element (index = 1) and generates items of the kind ', element'. This approach avoids having to check repeatedly within the loop whether we've reached the end or not in order to decide whether to include a comma or not.

`TListItem` has a helper method called `elementToString` which is used by `listToString` to print out a particular list item. This is shown below:

```
function TListItem.elementToString : string;
begin
  result := '';
  case self.itemType of
    liInteger : result := result + inttostr (self.iValue);
    liBoolean : result := result + boolToString (self.bValue);
    liDouble  : result := result + floattostr (self.dValue);
    liString  : result := result + self.sValue;
    liList    : result := result + (self.lValue).listToString;
  else
    raise Exception.Create('Unknown type in elementToString');
  end;
end;
```

We'll be adding other methods to `TListObject` as we go on.

8.3 New Opcodes

To support lists we only need three new opcode instructions:

```
   createList
   lvecIdx
   svecIdx
```

`createList` says what it does, it creates a list. It expects the elements of the list to be on the stack. The opcode takes an integer argument, which is the number of elements it expects to find on the stack. An example of how it would be called is given below:

```
# To create the list {4,6,9}
   pushi 4
   pushi 6
   pushi 8
   createList 3
```

In the `createList` method we have the following code:

```
function TVM.createList (count : integer) " TListObject;
var i : integer;
    p : PMachineStackRecord;
begin
  result := TListObject.Create (count);
  // Assume its garbage until it's assigned to a variable
  result.blockType := btGarbage;

  for i := count - 1 downto 0 do
      begin
      p := pop;
      case p.stackType of
        stInteger :
           begin
           result[i].iValue := p.iValue;
           result[i].itemType := liInteger;
           end;
       .. including Booleans and float ..
       stString:
          begin
          result.list[i].sValue := p.sValue.clone;
          result.list[i].sValue.blockType := btOwned;
          result.list[i].itemType := liString;
          end;
        stList:
          begin
          result.list[i].lValue := p.lValue.clone;
          result.list[i].lValue.blockType := btOwned;
          result.list[i].itemType := liList;
          end
      else
        raise Exception.Create('Unrecognized type in createList');
      end;
      end;
end;
```

One point worth mentioning is that the `for` loop for creating the list counts backward because the elements on the stack are in reverse order. To accommodate this approach, we must precreate an empty list of suitable length. This is achieved by the line: `TListObject.Create (count)`.

I've only shown code that deals with integers, string and lists. Double and Booleans are treated in the same way as an integer. Strings and lists are treated differently. For these objects, we clone them and set their memory status to `btOwned` since they are owned by the outer list.

This code takes care of situations like:

```
a = {3,4};
b = {1,2,a};
println (b);
```

```
{1,2,{3,4}}
c = b[2];
println (c);
{3,4}
```

When creating a list, if we find that a list subelement is bound to a variable, we clone the list as assign the clone to the subelement. If a list subelement is owned by another list, we also clone it before assigning it to the subelement. Once the list has been populated, we return the list to the caller. Changes, therefore, made to `b` will not be reflected by changes to `a`.

8.3.1 Adding Lists

As mentioned before, lists can be added:

```
a = {1,2} + {3,4}
```

This one of those operations that will leave behind garbage. `TListObject` has a class method called `addLists` shown below:

```
class function TListObject.addLists (list1, list2 : TListObject) : TListObject;
var i : integer;
begin
  if list1.blockType = btGarbage then
     result := list1
  else
     result := list1.clone;

  for i := 0 to list2.list.Count - 1 do
      result.list.Add (TListItem.Create (list2.list[i]));
end;
```

When adding lists together, we have to create a new list that corresponds to the sum. We can create this list from list1, the first argument in the method call. If the list is garbage, then we can use this directly, otherwise, we clone it first. After the sum, the memory map looks like:

```
GGB
```

The two garbage nodes correspond to the literal lists `{1, 2}` and `{3, 4}`. For a summation such as:

```
a = {1,2}
b = a + {3,4}
```

we get the memory map:

```
BGB
```

Only one garbage node is created, and a clone of the list in `a` is created for the sum. Consider adding multiple lists that only involve variables, for example summing three variables, such as:

```
a = {1,2}
b = {3,4}
c = {5,6}
>> d = a + b + c
```

This yields the following memory map where no garbage has been created.

```
BBBB
```

The bytecode used for this operation is shown below[1], where we see an intermediate list created by adding `a` to `b` before being added to `c`.

```
0  load 0
1  load 1
2  add
3  load 2
4  add
5  store 3
6  halt
```

8.3.2 Multiplying Lists by an Integer

Multiplying lists is similar to adding lists. A class method called `multiply` is responsible for multiplying lists. This method takes two arguments, an integer, and a list. The method creates an empty list that we repeatedly add the incoming list to. We return the list to the caller.

```
class function TListObject.multiply (value : integer;
                    aList : TListObject) : TListObject;
var i, j : integer;
    nContents : integer;
    workingCopy : TListObject;
    item : TListItem;
begin
  if aList.isBound then
     workingCopy := alist.clone
  else
```

[1]You can see the generated bytecode for any expression but first typing **debug** then executing the source code.

```
      workingCopy := aList;

   nContents := aList.list.Count;
   result := TListObject.Create(0);
   for i := 0 to value - 1 do
       if nContents = 0 then
          result.list.Add(TListItem.Create (0))
       else
          for j := 0 to nContents - 1 do
              begin
              item := TListItem.Create(workingCopy.list[j]);
              result.list.Add(item);
              end;
end;
```

8.3.3 Opcodes: `oLvecIdx` and `oSvecIdx`

The other two new opcodes, `oLvecIdx`, and `oSvecIdx`, are the load and store instructions when **indexing lists**. They will also be used when indexing strings.

`oLvecIdx` expects the stack to contain two items. The first item to pop off the stack will be index. The second item to pop off the list will be the list itself (Figure 8.1). `oLvecIdx` will use this information to extract the item from the list and push the item onto the stack.

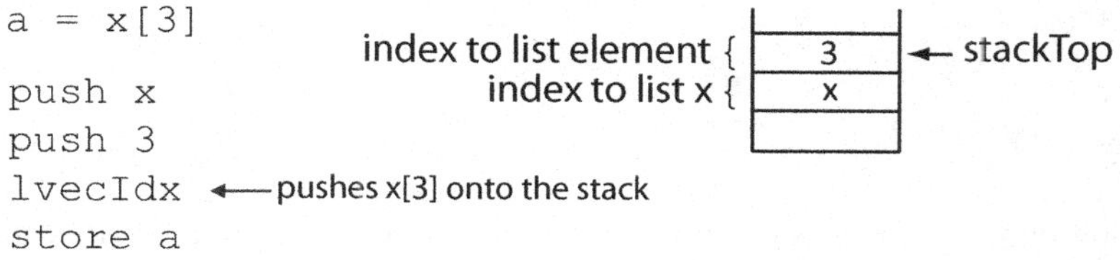

Figure 8.1 Indexing values using the syntax x[i] via the opcode `oLvecIdx`.

We have one small issue to also take care of. What happens when we write code where we have deeply nested lists, and we want to use multiple indexing? For example:

```
a = {1,2,{3,{4,5}}};
print (a[2,1])
{4,5}
```

There are different ways we can deal with this. We could push on to the stack the number of indices to expect, for example with `a[2,1,0]` we would use:

```
  pushi 0
  pushi 1
```

```
pushi 2
pushi 3
load a
lvecIdx
```

The alternative is to emit multiple `lvecIndx` instructions. This will work given the semantics for `oLvecIdx`. Each time we find an index we issue a `oLvecIdx` instruction. Using the previous example of `a[2,1,0]` we would write:

```
load 0
pushi 2
lvecIdx
pushi 1
lvecIdx
pushi 0
lvecIdx
```

This first loads variable `a` onto the stack, followed by the first index, 2. This is the order of operands that `oLvecIdx` expects to find on the stack. When `oLvecIdx` executes, it pushes the indexed item back on to the stack. If we push a second index onto the stack, we are now in a position to apply `oLvecIdx` again, this can be repeated as many times as there are indexes to apply.

A similar strategy can be applied when we wish to make an assignment such as:

```
a = {1,2,{3,{4,5}}};
a[2,1,0] = 99;
println (a)
{1,2,{3,{99,5}}}
```

However, one change has to be made to the last op code issued because we are doing an assignment. Instead of `oLvecIdx` we mush issue `oSvecIdx`. This leads to the following code:

```
pushi 99
load 0
pushi 2
lvecIdx
pushi 1
lvecIdx
pushi 0
svecIdx
```

What is nice about this is that whether we're assigning or accessing an indexed item, the code is the same except for the last op code. The other nice thing about this code is that we can also use it to handle strings. The code that handles `oLvecIdx` is shown below:

```
// Stack Structure to access indexed lists or strings such as [1]:
// index into symbol table containing indexable object
// Index into indexable object
// Called by oLvecIdx
procedure TVM.loadIndexable;
var
  index : integer;
  variable : PMachineStackRecord;
begin
  index := popInteger;
  variable := pop;
  case variable.stackType of
     stList   : loadIndexableList (variable, index);
    stString : loadIndexableString (variable, index);
  else
    raise ERuntimeException.Create('variable is not indexable ');
  end;
end;
```

This code also handles strings as well as lists using the methods: `loadIndexableList` and `loadIndexableString`. `loadIndexableList` is shown below:

```
procedure TVM.loadIndexableList(variable: PMachineStackRecord; index: integer);
var element: TListItem;
begin
  if (index < 0) or (index > variable.lValue.list.count - 1) then
      raise ERuntimeException.Create('list index out of range');

  element := variable.lValue.list[index];

  case element.itemType of
    liInteger: push(element.iValue);
    liBoolean: push(element.bValue);
    liDouble:  push(element.dValue);
    liString:  push(element.sValue);
    liList:    push(element.lValue);
  else
    raise ERuntimeException.Create ('internal error:
                 unsupported type in loadIndexableList');
  end;

end;
```

We do an initial check to make sure the index within bounds, access the indexed item, and push the item data type onto the stack.

`oSvecIdx` does the opposite to `oLvecIdx`, it will store a value on the stack into a list at a specific indexed position. `oSvecIdx` expects the stack to contain three items. The first item will be the index, the second item the symbol associated with the index, and the third item the value we wish to store in the indexable item (a list or a string). Depending on the data type of the indexable item we either call a method to deal with a string or a list.

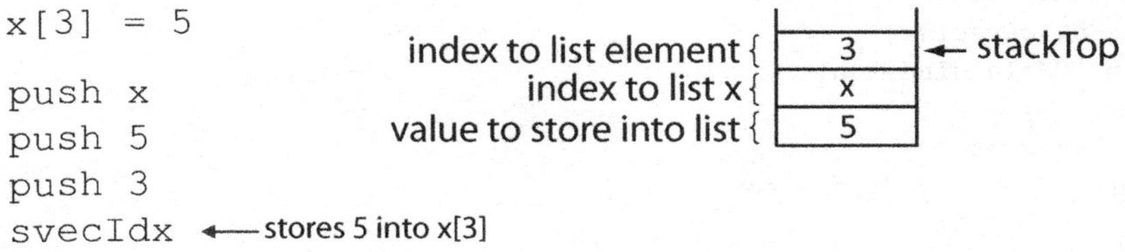

Figure 8.2 Storing values in indexed variables via the opcode `oSvecIdx`.

The code for `oSvecIdx` this is shown below:

```
// stack contains three items:
// 1) The index to the list or string
// 2) The indexable item, a list or string
// 3) The data to store into the indexable item,
procedure TVM.storeIndexable;
var
  index : integer;
  sym: PMachineStackRecord;
begin
  index := popInteger;
  sym := pop;
  if sym.stackType = stList then
    storeIndexableList(sym, index)
  else
  if sym.stackType = stString then
    storeIndexableString(sym, index)
  else
    raise Exception.Create('variable is not indexable');
end;
```

We can also define the equivalent `loadlocal`, and `storelocal` will handle access to any local lists or strings stored on the stack.

8.4 Running the Garbage Collector

One question that hasn't been addressed in when should garbage collection take place? This needs more thought than it might seem at first. Consider adding two literal lists:

```
a = {1,2} + {3,4}
```

Before the assignment is made, there are two lists with garbage status. If we were to garbage collect before the assignment is completed, there will be no list to assign to the variable a. In other words, we can't just garbage collect any time we like. The best time to collect garbage is after a store machine instruction has been executed. At that point, there shouldn't be any garbage nodes that are waiting to be assigned. Hopefully, there aren't any edge cases that I've not considered. This isn't very sophisticated, but it will do for now. For example, the store code can be modified as shown below:

```
oStore: begin
        storeSymbol(byteCode.index);
        if getGarbageSize() > garbageThreshold then
           collectGarbage; // Only collect garbage after a store
        end;
```

Rather than collect garbage every time we store a value, we can decide to do it less frequently. For example, as we create and free garbage we can keep a count of the size of the garbage pool. If the number of garbage nodes reaches a certain limit, say `garbageThreshold`, we initiate garbage collection.

`collectGarbage` just calls the `freeGarbage` method on the `memoryList` which is where references to all memory allocations are kept.

Bob Nystrom, in his online series on crafting an interpreter, has an interesting section called `When to Collect`[2]. There he discusses the pros and cons of throughput versus latency. Whereas the Rhodus collector in version I has a fixed interval for calling the garbage collector, Nystrom uses a threshold device where after a certain amount of allocation has occurred, the garbage collector is called. The other point he makes is that each time we garbage collect, we'll be checking nodes that don't need to be collected and may depending on the longevity be repeatedly checking them. In those cases, it might be worthwhile moving long-lived nodes to a sperate list of long-lived allocations. This list could be checked less frequently.

[2] https://craftinginterpreters.com/garbage-collection.html

9
Built-in Functions

9.1 Introduction

This will be a short chapter on adding built-in functions. From a user perspective calling a user-defined or built-in function is identical. Parsing user-defined or built-in functions should, therefore, also be very similar. The only difference is how they will be called by the virtual machine. To accommodate the difference, we'll add a new opcode to handle built-ins: `oBuiltin`.

But first, let's consider the parsing stage. When we parse functions, and we can't find the function name in the symbol table, we will check a separate list of built-in function names. If we find the built-in function name in the list, we proceed to parse the rest of the function call. Once a function name has been identified, be it a user function or a built-in, parsing is identical. For this reason, let's create a common method to parse user-defined and built-in functions. This method is shown below:

```
procedure TSyntaxAnalysis.parseFunctionCall (aProgram : TProgram;
                                  expectedNumArguments : integer);
var nArguments : integer; identifier : string;
begin
  identifier := sc.tokenString; nArguments := 0;
  sc.nextToken;
  if sc.token <> tRightParenthesis then
     nArguments := expressionList(aProgram);
  if nArguments <> expectedNumArguments then
     raise ESyntaxException.Create('incorrect number
          of arguments in function call: [' + identifier + ']');
  expect(tRightParenthesis);
end;
```

If we detect a user-defined function, we will call `parseFunctionCall` then emit the function index followed by the opcode `oCall`. On the other hand, if we find a built-in function name, we call `parseFunctionCall` and emit the built-in function index followed by the alternative opcode `oBuiltin`. That's all there is for parsing built-in or user-defined functions.

In the virtual machine we'll intercept `oBuiltin` and `oCall`:

```
oCall:     callUserFunction;
oBuiltin: callBuiltIn;
```

We've dealt with `callUserFunction` in a previous chapter, and there was quite a bit going on in that method. Thankfully, `callBuiltIn` is much simpler. With a built-in function, we don't have to deal with local variables, and arguments we pass to built-in functions will be much easier to handle. Just to show you how simple the method is, here it is:

```
procedure TVM.callBuiltIn;
var
  funcRecord: TBuiltinFunctionRecord;
  index: integer;
begin
  index := popInteger;

  funcRecord := builtinList[index];
  funcRecord.funcPtr(self);
end;
```

That's it. This is much shorter than `callUserFunction`. The only thing to describe is the `builtinList` variable, but it is but equally simple. We'll define the builtinList as:

```
TBuiltinList = class (TList<TBuiltInFunctionRecord>)
  public
   function find(const name: string; var index : integer): boolean;
end;
```

`builtinList` is a list of `TBuiltInFunctionRecord` types with an extra `find` method. The find method is what is called to identify a built-in method name. The built-in function record is given by:

```
TBuiltinFunctionRecord = record
  name : string;
  nArguments : integer;
  helpStr : string;
```

```
  funcPtr : TBuiltInFunction;
end;
```

Each built-in function will be associated with one of these records. This includes the name of the built-in, the number of arguments it takes, a help string, and a function pointer to the built-in itself.

We can create a helper routine to add entries to the builtinList, for example:

```
procedure addBuiltIns (name : string; nArguments : integer;
         helpStr : string; fcn : TBuiltInFunction);
var builtin : TBuiltInFunctionRecord;
begin
  builtin.name := name;
  builtin.nArguments := nArguments;
  builtin.helpStr := helpStr;
  builtin.funcPtr := fcn;
  builtinList.Add (builtin);
end;
```

An example of a built-in function is `int(x)`. This takes a number and returns the integer version of it. As with the Python equivalent, `int()` will truncate towards zero. Thus 3.4 is returned as 3, but so is 3.9. It essentially extracts the integer portion of a number. Passing an integer to the function simply returns the integer.

The function pointer, `TBuiltInFunction` has a specific signature:

```
  TBuiltInFunction = procedure (vm : TVM);
```

It expects a single argument which must be a reference to a virtual machine. The reason is that a built-in function needs access to the evaluation stack because this is where any arguments will be kept. As an example, the `int()` method is shown below:

```
procedure myInt (vm : TVM);
var value : PMachineStackRecord;
begin
  value := vm.pop;
  case x.stackType of
       stInteger : vm.push (value.iValue);
       stDouble  : vm.push (integer (trunc (value.dValue)));
  else
     argMustBeNumber;
  end;
end;
```

There isn't much to a built-in function. It first has to pop off any arguments it needs. It then carries out whatever operation it is expected to do and finally pushes any result on to the stack and exit. In the case of `int()` we make a check for what type the argument is. `int()` can only operate on integers (somewhat redundantly) or doubles. Anything else is treated as an error. To add `int()` to the `builtinList` we just call the following:

```
addBuiltIns ('int', 1, 'convert float to integer: int (3.4)', myInt);
```

```
>> println (int (3.9))
3
>> println (int (3.1))
3
>>
```

To make the current version of Rhodus more interesting, I also decided to add two input functions, one to read a string and another to read a number, which can be an integer or float.

```
readString()    // Read a string from the user
readNumber()    // Read an integer or float from the user
>>
```

The `readString` built-in method is straightforward. Input from the user is read using `readln`. This is used to create a string object which is pushed onto the stack before returning.

```
procedure readString (vm : TVM);
var s : string;
begin
  readln(s);
  vm.push (TStringObject.create (s));
end;
```

`readNumber` is a little bit more complicated because we have to convert a string input to the number, deal with users entering non-numbers, and distinguish between integers and floats. I used the `TryStrToInt` and `TryStrToFloat` to do the conversion because these don't cause an exception if the string is a badly formatted number. Instead, they return true or false. This makes it easier to handle bad input because I can create a while loop to repeatedly request input until the correctly formatted number is entered.

```
procedure readNumber (vm : TVM);
var s : string;
    iValue : integer; dValue : double;
begin
  readln(s);
  // No decimal point, then assume it could be a float
```

```
  if pos ('.', s) = 0 then
     begin
     if not TryStrToInt (s, iValue) then
        begin
        writeln ('Number error: ' + s + ' is not a number, try again');
        readln (s);
        while not TryStrToInt(s, iValue) do
           readln (s);
        end;
     vm.push (iValue);
     end
  else
     // Could be a float
     begin
     if not TryStrToFloat(s, dValue) then
        begin
        writeln ('Number error: ' + s + ' is not a number, try again');
        readln (s);
        while not TryStrToFloat(s, dValue) do
           readln (s);
        end;
     vm.push (dValue);
     end;
end;
```

The following illustrates these functions in operation:

```
print ("Enter a value to find the square root: ");
value = readNumber();
println ("The square root of ", value, " is ", value^0.5);
```

Running this script yields:

```
Enter a value to find the square root: hello
Number error: hello is not a number, try again
45
The square root of 45 is 6.70820393249937
```

Or using `readString()`:

```
print ("Enter the name of your favorite color: ");
str = readString();
println ("My favorite color is ", str);
```

Running this script yields:

```
Enter the name of your favorite color: blue
My favorite color is blue
```

Function	Description
`int (x)`	Returns the integer of the argument. The argument can be an integer or float value.
`readString()`	Returns a string from the console.
`readNumber()`	Returns an integer or float from the console.

Table 9.1 List of built-in functions.

That completes the built-in functions. Table 9.1 lists the current built-in functions.

10

A Glimpse at Part 3

10.1 Future Developments

User Visible Changes

In this chapter, we'll take a glimpse at what is coming in Part 3 of the series. The major addition will be modules. Modules will enable us to develop code libraries, both user-defined and built-in libraries. For example:

```
//Help: A library that provides routines to compute combinations and permutations
module mystuff

import math

//Help: Returns the number of permutations of n things taken r at a time
function permutations (n, r)
  return math.fact (n)/math.fact (n-r)
end;

//Help: Returns the number of combinations of n things taken r at a time
function combinations (n, r)
   return math.fact (n)/(math.fact(n-r)*math.fact (r))
end;

end
```

Some self-documenting help will be supported to allow users to request help at the interactive prompt. Modules can be imported and made available using the `import` keyword. The following modules will be developed in Part 3:

- Strings
- Lists
- Input/Output
- Math
- Memory inspection, garbage collection
- Time (for timing code)
- os (miscellaneous OS system functions)

A new data type will be introduced to support arrays. These will be n-dimensional collections of data stored contiguously in memory. A key feature of arrays is that they can only contain one data type per declaration. For example, an array of integers, an array of doubles, or an array of strings. This is in contrast to lists that are heterogeneous collections of data.

Exception Handling will be added as a means to trap and propagate runtime errors.

Internal Changes

The major internal change will be to separate syntax analysis from code generation and implement a proper AST phase.

A

List of Virtual Machine Opcodes

Abbreviation: TOS = Top Of Stack. The virtual machine stack grows upwards, therefore if the top of the stack is TOS, the next entry in the stack will be TOS-1.

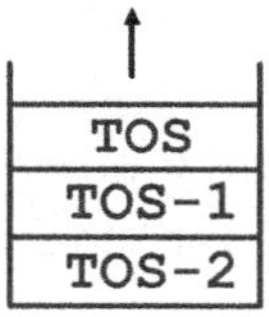

Figure A.1 Operand Stack: The stack grows upwards, TOS = Top Of Stack.

`Nop`

No operation

Description: This instruction performs no operation and can be used as a place holder.

Arithmetic Operations

`add`

Add operation

Description: This instruction will carry out addition by popping two values from the stack, adding their values and pushing the result back onto the stack.

`TOS = TOS-1 + TOS`

`sub`

Subtract operation

Description: This instruction will carry out subtraction by popping two values from the stack, subtracting their values, and pushing the result back onto the stack. The order of operands is given by:

pop value_1, pop value_2, push (value_2 - value_1)

`TOS = TOS-1 - TOS`

`mult`

Multiplication operation

Description: This instruction will carry out multiplication by popping two values from the stack, multiplying their values, and pushing the result back onto the stack. The order of operands is given by:

`TOS = TOS-1 × TOS`

`div`

Division operation

Description: This instruction will carry out division by popping two values from the stack, dividing their values, and pushing the result back onto the stack. The order of operands is given by:

pop value_1, pop value_2, push (value_2 / value_1)

`TOS = TOS-1 / TOS`

`mod`

Modulus operation

Description: This instruction will pop two items from the stack, compute the modulus then pushes the result back onto the stack

pop value_1, pop value_2, push (value_2 mod value_1)

`TOS = TOS-1 mod TOS`

`pow`

Power operation

Description: This instruction will compute the power by popping two values from the stack, computing one to the power of the other, and pushing the result back onto the stack. The order of operands is given by:

pop $value_1$, pop $value_2$, push ($value_2$^ $value_1$)

`TOS = POWER (TOS-1, TOS)`

`umi`

Unary-minus operation

Description: This instruction will pop off an item from the stack, negate it then push the result back onto the stack.

`TOS = -TOS`

`inc`

Increment symbol

Description: This instruction will use the value of the index value to access a value in the symbol table and increment the value by one.

`dec`

Increment symbol

Description: This instruction will use the value of the index value to access a value in the symbol table and decrement the value by one.

`localinc`

Increment symbol

Description: This instruction will use the value of the index value to access a value in a user function and increment the value by one.

`localdec`

Increment symbol

Description: This instruction will use the value of the index value to access a value in a user function and decrement the value by one.

Boolean Operations

`and`

AND operation

Description: This instruction will pop two items off the stack, compute the Boolean AND operation, then push the result back onto the stack.

`TOS = TOS-1 and TOS`

or

OR operation

Description: This instruction will pop two items off the stack, compute the Boolean OR operation, then push the result back onto the stack.

`TOS = TOS-1 or TOS`

not

NOT operation

Description: This instruction will pop off an item from the stack, compute the Boolean NOT operation, then push the result back onto the stack.

`TOS = not TOS`

xor

XOR operation

Description: This instruction will pop two items off the stack, compute the Boolean XOR operation, then push the result back onto the stack.

`TOS = TOS-1 xor TOS`

Push and Pop Instructions

pushi

---

push integer:

pushi value

Description: This instruction will push the integer value of the argument in the bytecode onto the stack.

pushb

---

push Boolean

pushb value

Description: This instruction will push the Boolean value of the argument in the bytecode onto the stack.

pushd

---

push double value

`pushd index`

Description: This instruction will take the integer argument in the bytecode and use it to index a floating point value in the constant table, which it then pushes onto the stack.

pushs

push string

`pushs index`

Description: This instruction will take the integer argument in the bytecode and use it to index a string stored in the constant table, which it then pushes onto the stack.

pop

pop a value off the stack

Description: This instruction will pop one item off the stack and return the value to the caller.

Branching Instructions

jmp

Unconditional jmp to a new instruction

Description: This instruction will use the bytecode integer argument as a relative jump distance.

jmpIfTrue

Jmp is the value on the stack is true

Description: This instruction will pop one item of the stack. If the popped item is True, the instruction will do a relative jump based on the bytecode integer argument.

jmpIfFalse

Jmp is the value on the stack is true

Description: This instruction will pop one item of the stack. If the popped item is False, the instruction will do a relative jump based on the bytecode integer argument.

User and Built-in Function Calls

`call`

Call a user-defined function.

Description: This instruction will call a user-defined function. The stack will contain the id index of the function and any optional arguments. When a call is made, the first item popped off the stack will be an integer index to the user function. The following items on the stack will be the arguments required by the user function. The first argument on the stack is the last argument in the function call.

`callBuiltin`

Call a builtin function.

Description: This instruction will call a built-in function. The stack will contain the id index of the built-in function and any optional arguments. When a call is made, the first item popped off the stack will be an integer index to the built-in function. The following items on the stack will be any arguments required by the built-in function. The first argument on the stack is the last argument in the built-in function call.

`ret`

Return from a function with a value

Description: Push a return value onto the stack and return from a call.

Symbol Table Instructions

`store`

Store a value to the symbol table

Description: Pop a value from the stack and copy the value to the symbol table. The index argument represents the index to the symbol table entry.

`load`

Load a value from the symbol table

Description: Load a value from the symbol table and push the value onto the stack. The index argument represents the index to the symbol table entry.

`localStore`

Store a value to the local storage area on the stack

Description: Pop a value from the stack and copy the value to local storage in a user function. The index argument represents the index to the local space in the user function.

`localLoad`

Load a value from the local storage area on the stack

Description: Load a value from the local storage in a user function and push the value onto the stack. The index argument represents the index to the local space in the user function.

Boolean Test Instructions

`isEq`

Test for Equality

Description: Pop two values from the stack, if they are equal in value, push True on the stack, else push False.

If `TOS-1 = TOS` then push true else push false

`isGt`

Test for greater than

Description: Pop two values from the stack, if one is greater than the other, push True on the stack, else push False.

If `TOS-1 > TOS` then push True else push False

`isGte`

Test for greater than or equal to

Description: Pop two values from the stack, if one is greater than or equal to the other, push True on the stack, else push False.

If `TOS-1 >= TOS` then push Frue else push False

`isLt`

Test for less than

Description: Pop two values from the stack, if one is less than the other, push True on the stack, else push False.

If `TOS-1 < TOS` then push True else push False

`isLte`

Test for less than or equal to

Description: Pop two values from the stack, if one is less than or equal to the other, push True on the stack, else push False.

If `TOS-1 <= TOS` then push True else push False

`isNotEq`

Test for not equal to

Description: Pop two values from the stack, if they are not equal to each other, push True on the stack, else push False.

If `TOS-1 <> TOS` then push True else push False

List Instructions

`createList`

Create a list: [1,2,3, etc]

Description: Use this instruction to create a list structure. The first item popped off the stack will be an integer representing the number of items on the stack to create the list.

The pseudo-code for this option is given below:

```
list = listObject
n = pop()
for i = 1 to n do
    list.append (pop())
end
```

`lvecIdx`

Load an indexed element of a list onto the stack

Description: Use this instruction to extract an element from a list, and push the value on to the stack. The first item on the stack will be the index, and the second item on the stack will be the list.

`svecIdx`

Store a value to an index element of a list

Description: Use this instruction to copy an item into a list. The first item on the stack will be the list, the second item the index into the list, and the third item the value to copy into the list.

`localLvecIdx`

Load an indexed element of a list onto the stack

Description: Use this instruction to extract an element from a list, and push the value on to the stack. The first item on the stack will be the list, and the second item on the stack the index into the list. Used to access lists that are defined locally in a user function.

`localSvecIdx`

Store a value to an index element of a list

Description: Use this instruction to copy an item into a list. The first item on the stack will be the list, the second item the index into the list, and the third item the value to copy into the list. Used to access lists that are defined locally in a user function.

Miscellaneous

`halt`

Halt the program

Description: Use this instruction to terminate the execution of a program.

`print`

Print a string

Description: Use this instruction to print one or more values on the stack. The instruction pops the first item from the stack, which will be the number of items to print. The method then issues a print event for each item out in turn (in reverse order). print events may be captured.

`println`

Print a string followed by a newline

Description: This is the same as `print` except right at the end it issues a newline character.

B

Interactive Console

A simple interactive console was written to make it easy to exercise Rhodus. The console code is nothing special at this stage, it's simplistic and full of unsupported edge cases, but it will do for now. At startup this is what the console looks like:

```
Welcome to Rhodus Language I Console, Version 1.0
Data and Time: 2/24/2020, 11:13:43 AM

quit             Quit the application
list fileName    List the contents of a file
run fileName     Run the code in the specified file
edit fileName    Start notepad to edit the file
dir              List the directory of .rh files
cls              Clear the screen
debug            Turn debugging on and off (displays bytecode)
mem              Report on current memory usage)
free             Run the memory garbage collector)
tests            Run the tests

?name            Get help about object name
??               List help topics

To run a script, just type its filename. Note: there is no need to specify the .rh extension
>>
```

When the console first runs it centers the screen before displaying the help screen. If we didn't do this, the console would open up at a different location every time, which I find particularly annoying. The console has a range of commands that include file handling, memory management, and a single command to run the regression tests.

I added some color to make the console look more interesting. The prompt is colored

red, and error messages are green. For those looking at the pdf, you see the following color scheme:

```
>> a = $
ERROR [line 1, column:  6] unrecongnised character in source code:$
>>
```

To implement color we'll provide these three helper routines:

```
procedure setRed;
begin
  setConsoleTextAttribute(conOut,
           FOREGROUND_INTENSITY or FOREGROUND_RED);
end;

procedure setGreen;
begin
  setConsoleTextAttribute(conOut,
         FOREGROUND_INTENSITY or FOREGROUND_GREEN);
end;

procedure setWhite;
begin
  setConsoleTextAttribute(conOut,
         FOREGROUND_INTENSITY or
         FOREGROUND_RED or
         FOREGROUND_GREEN or
         FOREGROUND_BLUE);
end;
```

To set the prompt color to red, we run the following script:

```
procedure displayPrompt;
begin
  setRed;
  try
    write ('>> ');
  finally
    setWhite;
  end;
end;
```

We do something similar for error messages, except they are printed in green. When the console starts, we need to call these three methods to allow us to change colors:

```
  setConsoleWindowPosition;
  conOut := getStdHandle(STD_OUTPUT_HANDLE);
  getConsoleScreenBufferInfo(ConOut, BufInfo);
```

B.1 Commands

The commands available at the console should be self-explanatory. The `debug` command flips a boolean flag so that whenever code is entered, it is dissembled first to the console before being run. Calling `debug` again flips the boolean flag off. There is some very primitive help that just lists the current set of built-in functions. `tests` let you run the regression tests to check everything is ok. These tests print out a dot if ok, or F is a particular test fails.

The `edit` command is worth saying something more about. During testing, I asked someone else to try it out. I noticed that each time they wanted to edit the same file, they would type `edit filename` each time. This resulted in multiple copies of Notepad being spawned. By the end of the session, there were 10s of copies of Notepad open, each displaying the same file. This was fairly annoying, so I added code that would first scan for all open Notepad windows, and if it found one that was already editing the same file, it would bring that window to the front, otherwise, it spawned a completely new Notepad window. The approach I took to implement this is not very conventional, but every Notepad window title has the form filename followed by `'- Notepad'`, for example: `'myprogram - Notepad'`. All I do is look for applications where the rightmost 10 characters in the windows title are : `'- Notepad'`. Each time I find one, I store it in a `winList` stringlist. After all desktop windows have been enumerated in this way, I check if any of the Notepad titles in `winList` contain the filename the user is asking to edit. If it finds the Notepad window for this particular file, it brings the window to the front, otherwise is spawns a new Notepad window.

The console application as it stands is probably not something you'd want to build on, and I'd recommend a rewrite if more functionality is to be added, for example, code completion and a history buffer. Eventually, there will be much more help support added to the compiler, and this will need to be handled better by the console.

With the current console, I've made it easy to run a script. Rhodus scripts should have the extension `.rh`. Type the name of the script without the extension will cause the script to be executed. For example, save the following script with the file name: `loopi.rh`

```
for i = 1 to 5 do
    println (i*i)
end
```

To create this file, type `edit loopi` at the console. The Windows notepad editor should appear asking if this is a new file, answer yes. Type in the script and select save. The `rh` extension will be automatically added when saved. Now type `loopi` at the console, and you'll see:

```
>> loopi
1
4
9
16
25
```

B.2 Minimal Code to Run a Script

This is the minimal code you need to run a script. The try/finally blocks that free the various objects are missing just to keep things simpler, but you'd need to add them in real code. We create the tokenizer, and pass that to the syntax analysis object. We create a main module that will hold the code and symbol table. We read in some source code, compile it into the mainModule, then pass the mainModule to the virtual machine. We register the print and println callbacks before running the module itself.

```
  sc := TScanner.Create;
  sy := TSyntaxAnalysis.Create (sc);
  mainModule := TModule.Create ('_main_');

  readln (sourceCode);
  sc.scanString(sourceCode);

  sc.nextToken;
  sy.mainProgram(mainModule);

  printObj := TPrintClass.Create;

  vm := TVM.Create;
  vm.registerPrintCallBack(printObj.print);
  vm.registerPrintlnCallBack(printObj.println);

  vm.runModule (mainModule);
```

The printObj is given below:

```
   TPrintClass = class (TObject)
      procedure print (st : PMachineStackRecord);
      procedure println (st : PMachineStackRecord);
   end;

procedure TPrintClass.print (st : PMachineStackRecord);
begin
  if st <> nil then
     case st.stackType of
```

```
        stNone     : begin end; //write ('undefined value'); end;
        stInteger : write (st.iValue);
        stDouble  : write (Format('%g', [st.dValue]));
        stString  : write (st.sValue.value);
        stBoolean : if st.bValue = True then
                       write ('True')
                    else
                       write ('False');
        stList    : begin
                    write (st.lValue.listToString);
                    end;
     else
        writeln ('Unrecognized value from print');
     end;
end;

procedure TPrintClass.println (st : PMachineStackRecord);
begin
  print (st);
  writeln;
end;
```

B.3 Interrupting a Running Program

There will be times when a program goes into an infinite loop or simply takes a long time to complete. For this reason, there needs to be a means to interrupt a running program. A common way to interrupt a program is for a user to hit CTRL-C. In a normally running console application, hitting CTRL-C will exit the program entirely, not exactly something we want to happen. In order to prevent this we need to trap CTRL-C.

The Windows operating system provides special handling of the CTRL-C key combination in console processes and is handled by a list of handler functions. By default this list only contains the default handler function which exits the application entirely. New handlers can be added programmatically using the method `SetConsoleCtrlHandler` so that an application can handle a CTRL-C in its own way. When a console receives a CTRL-C signal, the handler functions are called on a last-registered, first-called basis until one of the handlers returns True. If none of the handlers returns True, the default handler is called. With this in mind, let's create the following handler and register it with the operating system. This method can be added to the main program file.

```
function consoleHandler (dwCtrlType: DWORD) : BOOL; stdcall;
begin
  // Avoid terminating with Ctrl+C
  if (CTRL_C_EVENT = dwCtrlType ) then
```

```
    begin
    vm.stopVm();
    result := True
    end
  else
    result := False;
end;
```

To register this handler we just call the following method at startup:

```
setConsoleCtrlHandler (@consoleHandler, True);  // Ignore Ctrl-C
```

In the handler itself we check for CTRL-C; if detected, we return True to indicate we've handled this signal. In addition we call a new method in the virtual machine, called `stopVM()`. `stopVM()` is a simple affair and is shown below:

```
procedure TVM.stopvm;
begin
  bolStopVm := True;
end;
```

All this does is set a new private variable, called `bolStopVm`, in the `TVM` class to true. In the virtual machine instruction loop we add the following code inside the loop:

```
if bolStopVm then
   begin
   issueMessage ('Ctrl-C Interrupt Detected');
   exit;
   end;
```

I've also added a new method to issue messages from the virtual machine, in this case to issue an interrupt detected message. And that's it, a simple script to test this is:

```
   while True do
      println ('loop')
   end
```

Hitting CTRL-C will interrupt this infinite loop.

History

1. VERSION: 1.0

   Date: 2020-3-1

   Author(s): Herbert M. Sauro

   Title: Writing an Interpreter in Object Pascal: Part 2

   Modification(s): First Release to Printing Press

2. VERSION: 1.01

   Date: 2020-3-17

   Author(s): Herbert M. Sauro

   Title: Writing an Interpreter in Object Pascal: Part 2

   Modification(s): Updated Appendix B with CTRL-C discussion

Index

Made in United States
North Haven, CT
10 November 2026

22106112R00091